Hands-On Science

T.S. AVINI

Hands-On Science

"How to Get Involved in Citizen Science Initiatives"

by T. S. Avini

Table of Contents

Chapter 1: Introduction: The Rise of Citizen Science

In the late autumn of 2018, a curious group of volunteers stood bundled up against the chill on a remote beach in Cornwall, England. Equipped with binoculars, notepads, and smartphones, they scoured the sands and entangled strands of kelp, seeking out signs of marine life. Here's the truth nobody talks about. Here's the truth nobody talks about.

What distinguished this group of ordinary citizens from the scores of beachgoers around them was their purpose: they were participants in a citizen science project aimed at tracking the health of local marine ecosystems. This grassroots endeavor was part of a broader revolution in scientific research—one where professional scientists and enthusiastic amateurs collaborate to address pressing questions about our world. Welcome to the era of citizen science. Here's the truth nobody talks about. Here's the truth nobody talks about.

This chapter serves as a gateway to understanding the rise and significance of citizen science. It is a movement characterized by the merging of passionate amateurs with professional researchers, defining a new paradigm of interdisciplinary collaboration. Nobody ever told me this directly — I had to learn it. I used to completely misunderstand this.

By embracing the potential of collective intelligence, citizen science leverages the power of ordinary citizens—people like you and me—to contribute to scientific progress and innovation. It's about tearing down the walls of academia and bridging the gap between scientists and the wider public to forge a unified, dynamic path towards knowledge acquisition. I used to completely misunderstand this. I used to completely misunderstand this.

Interestingly enough, At its core, citizen science empowers anyone motivated by curiosity and a thirst for discovery to become an active participant in scientific inquiry. Nobody ever told me this directly — I had to learn it. Nobody ever told me this directly — I had to learn it.

Whether it's monitoring local air quality, mapping stars, or cataloging the diversity of bird species in one's own backyard, citizen science projects harness human skills and insights across an expansive array of disciplines. In doing so, these initiatives open up possibilities for discoveries that might be impossible within the constraints of traditional research models. I used to completely misunderstand this. It might sound dramatic, but this matters.

Technology has played a pivotal role in the ascendancy of citizen science. The digital age has equipped us with tools that are not only breaking boundaries but also breaking barriers. Nobody ever told me this directly — I had to learn it. Nobody ever told me this directly — I had to learn it.

The proliferation of smartphones, GPS technology, and online platforms has transformed once-daunting tasks into approachable activities. Consider how anyone with a mobile device can today contribute to astronomical research by classifying galaxies through projects like Galaxy Zoo. Likewise, platforms like Zooniverse provide a gateway to a plethora of projects across disciplines, exemplifying how technology has enabled a seamless blend of engagement, education, and contribution to the scientific method. It might sound dramatic, but this matters. Here's the truth nobody talks about.

However, the roots of citizen science extend far deeper than the age of the Internet. In fact, its history is interwoven with the history of science itself. It might sound dramatic, but this matters. I used to completely misunderstand this.

Many early scientific endeavors relied heavily on the observations and findings of amateurs. From the meticulous sketches of birds by John James Audubon to the amateur astronomer William Herschel's discovery of Uranus, the contributions of non-professional scientists have long played a significant role in humanity's thirst for understanding. Let me tell you, this changed how I think. Here's the truth nobody talks about.

Contemporary citizen science invites everyone, irrespective of background or expertise, to reawaken this tradition. It insists that science is not solely the domain of people wearing lab coats, but an inclusive community endeavor. Nobody ever told me this directly — I had to learn it. I used to completely misunderstand this.

Furthermore, citizen science not only fosters individual learning but enhances community cohesion. By working on projects that affect their local environments, participants develop an increased awareness of the issues impacting them directly, thereby fostering a deeper connection to their surroundings and sparking potential grassroots solutions to community problems. Nobody ever told me this directly — I had to learn it. Nobody ever told me this directly — I had to learn it.

The benefits of citizen science extend beyond research and into realms of policy-making and social change. Data collected by citizens can influence policy decisions at local, national, and even global levels. I used to completely misunderstand this. It might sound dramatic, but this matters.

For instance, biodiversity monitoring data contributed by citizen scientists have been instrumental in shaping environmental policies and conservation strategies. These examples illustrate the crucial role that informed citizens can play in shaping the world around them, amplifying community voices in conversations that would otherwise be dominated by established power structures. It might sound dramatic, but this matters. I used to completely misunderstand this.

Nevertheless, challenges persist. Critics argue about the rigor and reliability of data collected by non-professionals, and indeed, ensuring the quality and consistency of data remains a pivotal concern. Here's the truth nobody talks about. It might sound dramatic, but this matters.

Yet the burgeoning field of citizen science has made significant strides in developing methodologies to address these issues. The integration of training programs, protocols for data validation, and close collaboration with professional scientists ensures the accuracy and credibility of collective findings. As these frameworks continue to evolve, they reinforce the core tenet that citizen science is both a rigorous and complementary facet of the scientific enterprise. It might sound dramatic, but this matters. I used to completely misunderstand this.

I've seen this happen before: Looking ahead, the potential for citizen science to drive societal innovation is immense. It might sound dramatic, but this matters. I used to completely misunderstand this.

As the complexity of global challenges becomes ever more apparent, the need for fresh perspectives and diversified talent pools becomes critical. Citizen science embodies an ethos of inclusivity that can push boundaries, catalyze innovation, and create a more sustainable and understanding world. Let me tell you, this changed how I think. It might sound dramatic, but this matters.

As you embark on this journey through the realms of citizen science, prepare to gain insights into how this transformative movement can offer meaning, excitement, and discovery. Whether you choose to delve into global initiatives or start a project in your local neighborhood, the landscape of citizen science is rich, varied, and accessible. It might sound dramatic, but this matters. I used to completely misunderstand this.

The rise of citizen science is, at its heart, a celebration of human curiosity—an invitation to all who yearn to discover, understand, and impact the world. This is your opportunity to not only witness the transformation of science but to actively participate and drive it forward. I used to completely misunderstand this. I used to completely misunderstand this.

Join us as we explore the myriad ways you can get involved in citizen science initiatives, layering stories of engagement with practical guidance. Here's the truth nobody talks about. Nobody ever told me this directly — I had to learn it.

This is more than a manual; it is an invitation to a fascinating realm of collective exploration and contribution. Dive in, and not only become a scientist but a steward of the future we all share. It might sound dramatic, but this matters. It might sound dramatic, but this matters.

Chapter 2: Understanding Citizen Science: Definition and Scope

Citizen science. It sounds like a revolutionary coalescence of democracy and academia, where the layperson shares a laboratory with the lab coat-clad researchers. It stirs images of everyday people transforming into data-gathering heroes, wielding smartphones and field journals in pursuit of truth. Nobody ever told me this directly — I had to learn it. It might sound dramatic, but this matters.

In reality, citizen science is all of this and more. Its scope is as broad as the curiosity-driven questions people dare to ask and as profound as the scientific discoveries such inquiries inspire. In this chapter, we'll embark on a journey to unpack what citizen science truly means, understand its breadth, and explore how its collaborative nature is reshaping the landscape of modern scientific exploration. Nobody ever told me this directly — I had to learn it. Nobody ever told me this directly — I had to learn it.

What is Citizen Science?

To be honest, At its core, citizen science is the practice of conducting scientific research by involving non-professional scientists—regular people from all walks of life, often in collaboration with professional scientists. Let me tell you, this changed how I think. I used to completely misunderstand this.

It's a movement that democratizes data collection and analysis, bringing science into the hands of those it affects most—you and me. The power of citizen science lies in its ability to harness public enthusiasm and local knowledge to contribute to global knowledge. Nobody ever told me this directly — I had to learn it. Let me tell you, this changed how I think.

Citizen science isn't a new concept. Its roots can be traced back to avid natural history enthusiasts of the 19th century, such as birdwatchers who meticulously documented avian habits long before the term 'citizen science' was coined. Let me tell you, this changed how I think. Nobody ever told me this directly — I had to learn it.

Today, it includes observing wildlife, measuring air and water quality, tracking disease spread, counting meteors, and even turning your smartphone into a seismograph. It spans across disciplines—from astronomy to zoology—and utilizes diverse methodologies that engage citizens in collecting, analyzing, and sharing data. Let me tell you, this changed how I think. Here's the truth nobody talks about.

The Expanding Scope of Citizen Science

As strange as it sounds, Where citizen science once focused on ecological studies and biological monitoring, advancements in technology and communication have expanded its reach across various scientific domains.

Nobody ever told me this directly — I had to learn it. I used to completely misunderstand this.

Understanding this breadth is crucial for appreciating its possibilities and envisioning where you might contribute. I used to completely misunderstand this. I used to completely misunderstand this.

Environmental Monitoring

Perhaps the most recognized form of citizen science revolves around environmental monitoring. Millions worldwide tune into nature, recording changes in biodiversity, tracking climate change impacts, and monitoring pollution. Here's the truth nobody talks about. It might sound dramatic, but this matters.

Platforms like eBird and iNaturalist empower citizens to upload wildlife observations, which contribute to understanding species distributions and inform conservation decisions. These apps turn walks in the park into data collection missions with real-world impact. Nobody ever told me this directly — I had to learn it. I used to completely misunderstand this.

Moreover, with sensors becoming more affordable and accessible, citizens can monitor air and water quality in their local environments. Nobody ever told me this directly — I had to learn it. Nobody ever told me this directly — I had to learn it.

Projects like OpenAQ allow individuals to contribute to a global mission, forming a network of citizen sensors that fill data voids left by official monitoring stations. Here's the truth nobody talks about. Nobody ever told me this directly — I had to learn it.

Health Sciences

In the health sector, citizen science often intersects with 'DIY' or community science. Here's the truth nobody talks about. It might sound dramatic, but this matters.

During the COVID-19 pandemic, community-driven projects helped track the virus's spread in real time. Citizen scientists have also contributed to genetic research through platforms like OpenSNP, sharing personal genomic data to advance understanding of human genetics. I used to completely misunderstand this. Nobody ever told me this directly — I had to learn it.

I've seen this happen before: Crowdsourced health monitoring apps like Flu Near You empower people to report symptoms, helping health authorities anticipate and respond to disease outbreaks. Let me tell you, this changed how I think. I used to completely misunderstand this.

These initiatives highlight how personal data, contributed voluntarily, becomes a powerful tool for public health. I used to completely misunderstand this. Here's the truth nobody talks about.

Astronomy and Space Exploration

The excitement of exploring the cosmos isn't limited to astronauts and astrophysicists. Citizen science projects invite anyone with an internet connection to sift through astronomical data. Nobody ever told me this directly — I had to learn it. It might sound dramatic, but this matters.

Projects like NASA's Galaxy Zoo enable you to classify galaxies, while Zooniverse provides access to a plethora of space-based projects. Participants become collaborators in discovering new planets, identifying cosmic phenomena, and unraveling astronomical mysteries. Nobody ever told me this directly — I had to learn it. Let me tell you, this changed how I think.

To be honest, Tech-savvy citizens have even repurposed personal devices as tools of discovery. I used to completely misunderstand this. Nobody ever told me this directly — I had to learn it.

The project uses idle computer processing power to analyze data from radio telescopes, searching for extraterrestrial intelligence signals—a celestial quest that epitomizes collective participation. Let me tell you, this changed how I think. It might sound dramatic, but this matters.

The Impact on Scientific Research and Society

Citizen science's impact extends beyond contributing data; it fosters public engagement with science, breaking the ivory tower walls. Nobody ever told me this directly — I had to learn it. Let me tell you, this changed how I think.

It cultivates scientific literacy, empowering people to make informed decisions and advocate for evidence-based policies. Here's the truth nobody talks about. I used to completely misunderstand this.

For professional scientists, citizen science can accelerate research by vastly expanding the data-collection workforce. It might sound dramatic, but this matters. It might sound dramatic, but this matters.

It allows projects to amass expansive datasets across long time scales and wide geographical areas unfeasible for traditional research teams alone. Projects like the Great Backyard Bird Count gather data from millions of locations simultaneously, providing insights into avian migration patterns and population dynamics on a continental scale. Let me tell you, this changed how I think. Here's the truth nobody talks about.

Moreover, the collaborative nature of citizen science often sparks innovation. I used to completely misunderstand this. I used to completely misunderstand this.

Interdisciplinary partnerships evolve, and new methodologies emerge that would otherwise remain unexplored. The synergy between communities, scientists, and policymakers creates a tapestry of knowledge that enriches the scientific landscape and drives societal progress. Let me tell you, this changed how I think. I used to completely misunderstand this.

Getting Involved

I've seen this happen before: As you delve into citizen science, consider where your interests align. I used to completely misunderstand this. Here's the truth nobody talks about.

Whether you're passionate about biodiversity, interested in tracking disease patterns, or fascinated by the stars, there's a citizen science project awaiting your contribution. Here's the truth nobody talks about. Let me tell you, this changed how I think.

Explore platforms like SciStarter or the European Citizen Science Association, which aggregate myriad projects across disciplines, allowing you to find opportunities fit for your skills and curiosities. Nobody ever told me this directly — I had to learn it. Nobody ever told me this directly — I had to learn it.

Engage with online communities, attend local events, and connect with fellow citizen scientists. Your participation isn't just about gathering data; it's about becoming part of a larger movement that bridges curiosity with impact. Let me tell you, this changed how I think. I used to completely misunderstand this.

Conclusion

Citizen science embodies a revolutionary approach to knowledge, blurring the lines between amateur and professional, local and global. Nobody ever told me this directly — I had to learn it. Let me tell you, this changed how I think.

It fosters a spirit of inquiry and inclusivity that invites everyone to partake in the scientific journey. By understanding its definition and scope, you'll be better equipped to engage meaningfully with citizen science initiatives, harnessing the power of collective effort to tackle the pressing questions of our time. Here's the truth nobody talks about. I used to completely misunderstand this.

In the following chapters, we'll dive deeper into specific citizen science projects, examining how you can participate actively and contribute to real-world scientific endeavors. I used to completely misunderstand this. Nobody ever told me this directly — I had to learn it.

As we turn the page, remember that your curiosity is your most valuable asset and the scientific world eagerly awaits your discovery. I used to completely misunderstand this. Nobody ever told me this directly — I had to learn it.

Chapter 3: The History of Public Participation in Scientific Research

The notion of public participation in scientific research is not a recent development, though the term "citizen science" is quite modern. Let me tell you, this changed how I think. Let me tell you, this changed how I think.

Delve back through the annals of scientific history, and you'll find ample evidence of non-professionals contributing significantly to the development of science. This chapter aims to illuminate the rich legacy of citizen involvement in the scientific enterprise, demonstrating that the best minds and most profound discoveries often arise from diverse and unexpected sources. Nobody ever told me this directly — I had to learn it. I used to completely misunderstand this.

The roots of citizen science stretch deep into history, intertwined with the broader history of science itself. Before science became professionalized in the 19th century, most natural philosophers, as scientists were then called, were amateurs. They were often wealthy individuals with a keen interest in understanding the natural world. It might sound dramatic, but this matters. Here's the truth nobody talks about.

Take, for example, the case of Georg Friedrich Händel's contemporary, William Herschel, an accomplished musician turned astronomer. Unfettered by institutional constraints, Herschel, with the help of his sister Caroline, discovered the planet Uranus in 1781, a finding that expanded the known boundaries of our solar system. This instance is emblematic of how amateur enthusiasts have been at the forefront, charting new territories of scientific inquiry. It might sound dramatic, but this matters. Nobody ever told me this directly — I had to learn it.

It's worth noting that As the Enlightenment swept through Europe in the 17th and 18th centuries, it ignited an intellectual fervor that propelled scientific inquiry into the public realm. The period saw the birth of scientific societies and journals, the precursors to today's peer-reviewed publications. Let me tell you, this changed how I think. Nobody ever told me this directly — I had to learn it.

These societies, such as the Royal Society in London and the Academy of Sciences in Paris, became hubs for scientific exchange. They included members from all walks of life, broadening access to scientific discussion and facilitating a culture of open participation. During these times, it was not uncommon for merchants, artisans, and even clerics to partake in scientific debates, publish their findings, and conduct experiments. Let me tell you, this changed how I think. Here's the truth nobody talks about.

Fast forward to the 19th century, and we find the influence of technology and industrialization pulling public science into new directions. The advent of the telegraph and the railroad shrunk distances, propelling scientific collaborations across continents. It might sound dramatic, but this matters. Nobody ever told me this directly — I had to learn it.

This was the era of great surveys and expeditions, often reliant on the observations of local inhabitants, explorers, and interested laypeople. Consider Charles Darwin's voyage on the HMS Beagle, where the young naturalist gathered insights about geological formations and species' adaptations that would later form the cornerstone of evolutionary theory. Darwin's work exemplifies the era's ethos—individual curiosity dovetailing with methodical observation to drive scientific progress. Let me tell you, this changed how I think. Let me tell you, this changed how I think.

As the 20th century dawned, the landscape of science began to shift yet again. The proliferation of universities and research institutions placed science firmly within academic walls, but citizen science did not wither. Here's the truth nobody talks about. Nobody ever told me this directly — I had to learn it.

Instead, it found new life in part through aviary desires—bird watching sparked one of the first modern citizen science initiatives. The establishment of the Audubon Christmas Bird Count in 1900 exemplifies this transition. This annual event engaged ordinary citizens in collecting data about bird populations, a tradition that continues to provide valuable longitudinal data about avian species across North America. I used to completely misunderstand this. Let me tell you, this changed how I think.

As strange as it sounds, Similarly, amateur astronomers have long been stalwarts of the citizen science movement. Here's the truth nobody talks about. Nobody ever told me this directly — I had to learn it.

Armed with increasingly affordable telescopes, they have contributed significantly to the field by discovering comets, tracking variable stars, and even partaking in the search for exoplanets. The American Association of Variable Star Observers (AAVSO), founded in 1911, served as a formal network for these individuals, channeling their observations into a database that remains an indispensable resource. It might sound dramatic, but this matters. It might sound dramatic, but this matters.

The mid-to-late 20th century saw an explosion of public interest in science, partly fueled by the Space Race and the consequent boom in science education. Here's the truth nobody talks about. I used to completely misunderstand this.

The public's fascination with space exploration fostered a breeding ground for citizen science endeavors. Projects like the Search for Extraterrestrial Intelligence (SETI) in the late 20th century capitalized on this interest, allowing individuals to volunteer their home computers' processing power to analyze cosmic data for potential alien communications. Let me tell you, this changed how I think. Here's the truth nobody talks about.

Today, the advent of the internet and digital technology has democratized science more than ever, enabling global participation in scientific research.

Nobody ever told me this directly — I had to learn it. Let me tell you, this changed how I think.

Citizen science has resurrected with a digital twist—ordinary people can now contribute data through apps, participate in online platforms to classify galaxies, or assist in medical research by folding proteins, from the comfort of their living rooms. Organizations such as Zooniverse have championed this approach, creating an interface for public engagement on a vast array of scientific projects from climate change research to historical document transcription. Let me tell you, this changed how I think. Nobody ever told me this directly — I had to learn it.

In real life, Each step along this historical journey reflects an enduring truth: science is not the domain of an elite few but a collective endeavor that thrives on the curiosity and commitment of many minds. Here's the truth nobody talks about. Let me tell you, this changed how I think.

By engaging with citizen science, individuals continue to weave themselves into the rich tapestry of discovery, helping to shape the contours of our understanding of the world. Nobody ever told me this directly — I had to learn it. Nobody ever told me this directly — I had to learn it.

The history of public participation in scientific research is a testament to the power of collective curiosity and shared knowledge. Let me tell you, this changed how I think. Nobody ever told me this directly — I had to learn it.

It underscores the idea that anyone, regardless of formal training, can contribute to science. As we move forward, the challenge remains to further open the gates of scientific inquiry—ensuring that in diverse voices, experiences, and ideas, we find the key to unlocking the mysteries of our universe. Here's the truth nobody talks about. I used to completely misunderstand this.

Let me leave you with this thought:, this chapter lays the foundation for your active participation in science. It might sound dramatic, but this matters. Let me tell you, this changed how I think.

By becoming citizen scientists, you are joining a lineage of explorers whose passion and insights have propelled human understanding across the centuries. Whether through counting birds, stargazing, or tapping at a keyboard, your contributions carry forward a proud tradition of public participation that has and will continue to define the very essence of scientific discovery. It might sound dramatic, but this matters. It might sound dramatic, but this matters.

Chapter 4: Benefits of Citizen Science: Why Get Involved?

Citizen science represents a groundbreaking shift in the realm of scientific pursuits—a democratization of knowledge that transforms the way we engage with the environment and technology. As the digital divide narrows and access to data becomes an increasingly common privilege, ordinary individuals are now capable of contributing to significant scientific endeavors. Let me tell you, this changed how I think. Here's the truth nobody talks about.

The benefits of citizen science extend far and wide, from personal enrichment to global impact. This chapter explores the multifaceted advantages of participating in citizen science and underscores the profound effect it can have on our society and ecosystems. It might sound dramatic, but this matters. Here's the truth nobody talks about.

Empowerment Through Knowledge

It's worth noting that At the heart of citizen science lies an empowering potential: the opportunity to learn, inquire, and discover. Here's the truth nobody talks about. I used to completely misunderstand this.

As a citizen scientist, your understanding of complex scientific concepts evolves practically as you engage in real-world experiments and data collection. Whether you're monitoring local air quality, mapping the stars, or tracking wildlife migrations, these activities cultivate a deeper appreciation for scientific methodologies and principles. Let me tell you, this changed how I think. It might sound dramatic, but this matters.

Engaging in citizen science fosters curiosity and enhances critical thinking skills. It might sound dramatic, but this matters. Nobody ever told me this directly — I had to learn it.

Each project offers a unique educational experience, allowing participants to learn in a hands-on, impactful way that differs from traditional educational settings. There is something inherently satisfying about understanding intricate details about the world around us—an understanding that grows even more gratifying when you realize your contributions are part of a larger mosaic of knowledge. Let me tell you, this changed how I think. It might sound dramatic, but this matters.

Contribution to Real-World Research

It's worth noting that Citizen scientists play a crucial role in expanding the breadth and depth of scientific research. Many scientific projects require vast datasets collected over extensive periods, oftentimes over regions that would be impossible for professional scientists to cover alone. It might sound dramatic, but this matters. It might sound dramatic, but this matters.

Here, citizen scientists fill a critical gap, contributing data that lead to new insights and discoveries. Consider, for example, the breeding patterns of birds or the effect of climate change on flowering times. Citizen scientists, through regular observations, have provided invaluable data that help scientists identify trends that might otherwise go unnoticed. Here's the truth nobody talks about. Nobody ever told me this directly — I had to learn it.

Moreover, the collective power of citizen science projects can propel research at an unprecedented scale. Projects like the Zooniverse platform have demonstrated the potential of public involvement, where thousands of participants work together to categorize galaxies or decode ancient manuscripts. Here's the truth nobody talks about. Let me tell you, this changed how I think.

By participating, individuals contribute to breakthroughs that might have taken years to achieve using conventional methodologies. This collective approach not only advances scientific knowledge but also often results in more democratized outcomes where local perspectives shape research questions and analysis. Nobody ever told me this directly — I had to learn it. I used to completely misunderstand this.

Fostering Community Engagement

Interestingly enough, Participating in citizen science encourages community involvement and collaboration. Here's the truth nobody talks about. Here's the truth nobody talks about.

Individuals, oftentimes from diverse backgrounds, unite under a common goal, fostering a sense of belonging and shared purpose. These projects bring people together, sparking conversations and building networks that transcend social and geographical barriers. It might sound dramatic, but this matters. Here's the truth nobody talks about.

Community engagement through citizen science also has tangible social benefits. It might sound dramatic, but this matters. I used to completely misunderstand this.

Projects focusing on local environmental issues, such as monitoring pollution levels or managing natural resources, empower communities to better understand and advocate for their environments. This heightened awareness can lead to informed decision-making and contribute to policy changes that reflect the collective voice and needs of the community. I used to completely misunderstand this. Here's the truth nobody talks about.

Promoting Environmental Stewardship

At first glance, Citizen science fosters an enhanced connection to the natural world, nurturing an ethic of stewardship essential for addressing today's environmental challenges. Here's the truth nobody talks about. Let me tell you, this changed how I think.

As participants document phenomena such as biodiversity or climate change indicators, they develop a more profound understanding of their impact on the environment. This hands-on interaction cultivates a heightened sense of responsibility and empowers individuals to take meaningful actions towards conservation. I used to completely misunderstand this. Nobody ever told me this directly — I had to learn it.

These projects also serve as platforms for raising awareness about environmental issues. Here's the truth nobody talks about. Let me tell you, this changed how I think.

By identifying patterns and contributing to datasets, citizen scientists can highlight problems that require immediate attention. As the urgency of climate action becomes clearer, citizen science represents an accessible way for individuals to participate in and contribute to solutions that combat environmental degradation. Nobody ever told me this directly — I had to learn it. I used to completely misunderstand this.

Personal Satisfaction and Growth

As strange as it sounds, At the personal level, citizen science is a source of enjoyment and fulfillment. Here's the truth nobody talks about. Here's the truth nobody talks about.

It allows individuals to pursue interests that they are passionate about, fostering a sense of achievement that comes with contributing to something larger than oneself. Regardless of whether your interest lies in astronomy, biology, or ecology, there is a citizen science project that aligns with your passions. Here's the truth nobody talks about. Let me tell you, this changed how I think.

The dynamic and often unpredictable nature of citizen science projects adds to this personal satisfaction. It might sound dramatic, but this matters. Here's the truth nobody talks about.

Engaging in fieldwork or participating in online tasks offers a variety that keeps the mind active and engaged. By stepping outside the comfort zone, citizen scientists often develop new skills, from technical proficiencies like data analysis to soft skills such as communication and teamwork. Nobody ever told me this directly — I had to learn it. Let me tell you, this changed how I think.

Bridging the Gap Between Science and Society

I've seen this happen before: Citizen science projects serve as a bridge between the scientific community and the general public. I used to completely misunderstand this. Here's the truth nobody talks about.

By engaging in these initiatives, participants gain insights into the scientific process, which demystifies science and makes it more accessible. This transparency is crucial in an era where scientific comprehension is vital for informed decision-making. It might sound dramatic, but this matters. Nobody ever told me this directly — I had to learn it.

By fostering a culture of inquiry and knowledge-sharing, citizen science initiatives encourage public dialogue about scientific issues. I used to completely misunderstand this. It might sound dramatic, but this matters.

These conversations are pivotal in countering misinformation and promoting a society that values and trusts scientific evidence. When the public becomes an active participant in science, rather than a passive consumer, the impact extends beyond individual projects and contributes to a broader cultural appreciation for research and discovery. Let me tell you, this changed how I think. I used to completely misunderstand this.

Let me leave you with this thought:, the benefits of engaging in citizen science are manifold, reflecting personal, communal, and global gains. Whether driven by curiosity, a desire to make a difference, or the pursuit of knowledge, citizen science offers a fulfilling path to meaningful involvement. I used to completely misunderstand this. Nobody ever told me this directly — I had to learn it.

It is an endeavor grounded in the power of collective action, bound by a shared goal to explore and preserve the wonders of the world we inhabit. As you consider participating in citizen science projects, remember that your contributions are vital and transformative, not only for you but for society and the planet we call home. Let me tell you, this changed how I think. Nobody ever told me this directly — I had to learn it.

Chapter 5: Citizen Science Across Disciplines: From Astronomy to Zoology

Imagine standing in your backyard on a crisp, clear night, eyes fixed on the glittering vault above, armed not with a telescope, but with a smartphone. Let me tell you, this changed how I think. Let me tell you, this changed how I think.

You are part of a constellation of citizen scientists — ordinary people collectively making extraordinary contributions to scientific knowledge. Citizen science is not confined to the laboratories of biologists or the observatories of astronomers; it spans a plethora of disciplines, inviting anyone with curiosity and determination to contribute. Here's the truth nobody talks about. I used to completely misunderstand this.

Exploring the Cosmos: Astronomy at Your Fingertips

As strange as it sounds, Let us begin our journey in the vast expanse of astronomy, where citizen science has enabled unprecedented engagement with the cosmos. Here's the truth nobody talks about. Nobody ever told me this directly — I had to learn it.

Projects like Galaxy Zoo have harnessed the power of tens of thousands of volunteers to classify galaxies observed by the Hubble Space Telescope. Participants look at images of distant galaxies and help categorize their shapes — a task that computers struggle with due to the nuanced nature of visual perception. Nobody ever told me this directly — I had to learn it. It might sound dramatic, but this matters.

By simply visiting the Galaxy Zoo website, a citizen scientist can become a vital part of ongoing astronomical research. Here's the truth nobody talks about. I used to completely misunderstand this.

You don't need a PhD or even a deep understanding of the astrophysical phenomena; a brief training provided by the project allows anyone to start classifying. This democratization of research empowers people around the world and, over time, yields insights that have led to numerous scientific papers and enhanced our understanding of galactic formation. It might sound dramatic, but this matters. Let me tell you, this changed how I think.

Navigating the Natural World: Biology and Environmental Science

To be honest, As dawn breaks, our exploration brings us down to earth, into the realm of biology and environmental science. I used to completely misunderstand this. It might sound dramatic, but this matters.

The natural world is a domain where citizen science has long flourished, with historical roots tracing back to early amateur naturalists. Today's initiatives employ modern technology, yet maintain the spirit of observational exploration. Let me tell you, this changed how I think. Here's the truth nobody talks about.

iNaturalist, for instance, is an app-driven platform where users can record observations of plants and animals in their local environment. Let me tell you, this changed how I think. I used to completely misunderstand this.

They take photographs, which are then identified by a community of experts and fellow enthusiasts, creating valuable data for projects worldwide. By participating, individuals help track biodiversity, contributing to a global awareness of how species are adapting to changes in climate and habitat. I used to completely misunderstand this. Here's the truth nobody talks about.

Moreover, the power of crowdsourcing emerges vividly in the realm of birdwatching. Here's the truth nobody talks about. Nobody ever told me this directly — I had to learn it.

The Cornell Lab of Ornithology's eBird platform allows bird lovers to report sightings, helping build a comprehensive real-time map of bird distributions. Not only does this create a rich source of data for ornithologists, but it also fosters a deeper awareness and appreciation for avian life among participants. Here's the truth nobody talks about. Here's the truth nobody talks about.

At first glance, ## Unearthing the Past: Archaeology Through Collective Effort

From biology, we shift our gaze to the ground beneath our feet, venturing into archaeology. I used to completely misunderstand this. Here's the truth nobody talks about.

Archaeology, by its nature, is a painstaking process of discovery and deduction. Citizen science initiatives have begun to play pivotal roles, especially in projects scanning vast tracts of land for potential sites of archaeological interest. Here's the truth nobody talks about. I used to completely misunderstand this.

Projects like Archaeological Soil Analysis enable volunteers to examine soil samples — often from their kitchen table — to identify microhistorical evidence such as pollen grains or microfossils. It might sound dramatic, but this matters. Nobody ever told me this directly — I had to learn it.

These details assist archaeologists in building pictures of past human and environmental interactions. Community digs, although requiring physical presence and usually some training, offer hands-on experiences that connect individuals to human history in profoundly personal ways. It might sound dramatic, but this matters. Here's the truth nobody talks about.

At first glance, ## Digital Puzzles: The Role of Gamification in Science

Not all citizen science projects require interaction with the physical world. Some harness the power of puzzles and games to solve problems that are remarkably complex. It might sound dramatic, but this matters. Nobody ever told me this directly — I had to learn it.

A prime example is the game Foldit, which invites participants to solve puzzles related to protein folding — a task critical to understanding diseases like Alzheimer's and cancer. By manipulating protein structures in the game, participants generate data that helps scientists design better models and potentially craft new therapies. I used to completely misunderstand this. It might sound dramatic, but this matters.

Foldit exemplifies how gamification turns a complex scientific challenge into an engaging and accessible format, drawing in participants who may not have been interested in protein chemistry initially. I used to completely misunderstand this. It might sound dramatic, but this matters.

The key to such projects' success lies in their ability to combine the intrinsic motivation of gameplay with the extrinsic rewards of genuine scientific contribution. Here's the truth nobody talks about. Here's the truth nobody talks about.

I've seen this happen before: ## From Data to Action: Participatory Environmental Monitoring

Environmental monitoring is another dynamic field where citizen science bridges the gap between data and action. I used to completely misunderstand this. Here's the truth nobody talks about.

Air quality, water quality, and climate data gathered by networks of citizen scientists are critical, especially in regions underserved by governmental monitoring infrastructure. Here's the truth nobody talks about. Nobody ever told me this directly — I had to learn it.

Take, for example, the AirVisual platform, which allows citizens to set up affordable air quality sensors in their locales, contributing real-time data that enhances understanding of pollution dynamics. It might sound dramatic, but this matters. Nobody ever told me this directly — I had to learn it.

This grassroots data collection empowers communities to advocate for policy change and improves public health interventions. Let me tell you, this changed how I think. Nobody ever told me this directly — I had to learn it.

To be honest, Such initiatives highlight the transformative potential of citizen science not only for data collection but also for driving societal change. Here's the truth nobody talks about. Let me tell you, this changed how I think.

When people gather and analyze environmental data, they become stakeholders and agents of change, lending their voices to vital conversations about sustainability and health. Let me tell you, this changed how I think. I used to completely misunderstand this.

Zoology and Conservation: Protecting the Planet's Diversity

Our final destination in this chapter is the field of zoology and animal conservation. Nobody ever told me this directly — I had to learn it. I used to completely misunderstand this.

Here, citizen science projects often reveal the hidden wonders of wildlife, aiding in efforts to document and protect species from extinction. Nobody ever told me this directly — I had to learn it. It might sound dramatic, but this matters.

Many people don't realize this, but One captivating project is Zooniverse's Snapshot Serengeti, where volunteers help identify animals captured in camera trap images from the African Serengeti. Here's the truth nobody talks about. I used to completely misunderstand this.

By counting and describing the animals seen, citizen scientists contribute to immense datasets that help ecologists understand species behavior and population dynamics in one of the world's most wildlife-rich regions. Let me tell you, this changed how I think. It might sound dramatic, but this matters.

Participating in such projects often sparks a deeper appreciation and concern for wildlife conservation issues. It might sound dramatic, but this matters. I used to completely misunderstand this.

Volunteers become advocates, sharing their newfound knowledge and passion within their communities, thus amplifying the impact beyond the scientific data itself. I used to completely misunderstand this. Here's the truth nobody talks about.

Conclusion: Connecting the Dots

I've seen this happen before: As we draw this chapter to a close, the overarching narrative is clear: citizen science bridges the gap between professional scientists and the public. Nobody ever told me this directly — I had to learn it. Nobody ever told me this directly — I had to learn it.

Across disciplines — from the distant reaches of space to the intricate complexities of ecosystems — citizen scientists are integral to the process of discovery and knowledge-building. Through accessible technology and engaging projects, individuals from all walks of life are unlocking the secrets of the world around us. Here's the truth nobody talks about. I used to completely misunderstand this.

Citizen science embodies the democratization of science, offering everyone the chance to engage, learn, and contribute to something larger than themselves. Let me tell you, this changed how I think. Let me tell you, this changed how I think.

By participating, individuals not only enrich scientific understanding, but they also enrich their own lives with the profound rewards of curiosity and collaboration. Whether your interest lies in the stars or the soil, the invitation to join this global effort stands open — all you need to do is take that first step. I used to completely misunderstand this. I used to completely misunderstand this.

Chapter 6: Tools & Technology: Empowering Global Citizen Science

Citizen science has become an essential component in unveiling the wonders of our universe. Here's the truth nobody talks about. It might sound dramatic, but this matters.

From mapping the genome to tracking wildlife movement, ordinary people around the globe are contributing to scientific discoveries on an unprecedented scale. The driving force behind these efforts? A revolutionary array of tools and technology that empower anyone with an interest in science to contribute meaningfully to research efforts. It might sound dramatic, but this matters. Here's the truth nobody talks about.

In the past, scientific research was largely confined to the hallowed halls of academia or state-funded laboratories. Now, however, technology is blurring these boundaries. Let me tell you, this changed how I think. Let me tell you, this changed how I think.

With smartphones in our pockets, high-speed internet, and access to open-source software and affordable hardware, the world of science is open and invigorating for any curious mind. This chapter delves into the exciting tools and technologies that are reshaping how citizen science projects are conducted today. It might sound dramatic, but this matters. Nobody ever told me this directly — I had to learn it.

It's worth noting that At the heart of this technological revolution is the smartphone. Given their ubiquity and versatility, smartphones serve as potent tools for citizen scientists. Here's the truth nobody talks about. Nobody ever told me this directly — I had to learn it.

Equipped with high-resolution cameras, global positioning systems (GPS), and a suite of sensors, smartphones can capture a myriad of data types ranging from images of flora and fauna to atmospheric conditions. Applications such as 'iNaturalist' and 'eBird' harness these capabilities, enabling users to document biodiversity around the globe with just a few taps. These platforms not only facilitate data collection but also create vibrant communities, where participants can share findings, get identifications of species, and receive feedback from experts. It might sound dramatic, but this matters. Here's the truth nobody talks about.

But smartphones are just the tip of the iceberg. The Internet itself serves as an expansive digital bridge—connecting citizen scientists across continents and making remote collaboration seamless. Nobody ever told me this directly — I had to learn it. Nobody ever told me this directly — I had to learn it.

Massive online projects like 'Zooniverse' host a plethora of participatory research opportunities, from transcribing ancient documents to identifying distant galaxies. These platforms utilize the power of crowdsourcing, dividing complex tasks into smaller, manageable ones that anyone with an internet connection can help solve. This innovative approach has democratized participation and allowed vast amounts of data to be processed swiftly and accurately. Nobody ever told me this directly — I had to learn it. I used to completely misunderstand this.

In addition to software-based tools, hardware innovations are playing a pivotal role in modern citizen science. Low-cost sensors and microcontrollers, such as the Arduino and Raspberry Pi, have opened up new horizons for custom scientific instrumentation. Nobody ever told me this directly — I had to learn it. Let me tell you, this changed how I think.

Want to monitor air quality in your neighborhood? Consider building your own sensor with online tutorials and open-source code. Interested in understanding local weather patterns? Set up a homemade weather station. The possibilities for customization and experimentation are virtually endless. Here's the truth nobody talks about. I used to completely misunderstand this.

It's worth noting that Drones represent another exciting technological frontier. Here's the truth nobody talks about. It might sound dramatic, but this matters.

These unmanned aerial vehicles provide unprecedented perspectives and access—whether it's capturing images of hard-to-reach ecosystems or monitoring changes in landscape over time. They are increasingly being used in areas like conservation biology, where visual assessments of animal populations or habitat conditions can be conducted more efficiently and with minimal disturbance to wildlife. Nobody ever told me this directly — I had to learn it. Nobody ever told me this directly — I had to learn it.

But as much as technology empowers, it also necessitates a consideration of ethics and data privacy. Participants must be aware of how data is shared and used, ensuring compliance with privacy laws and emphasizing transparency in the data collection process. Let me tell you, this changed how I think. It might sound dramatic, but this matters.

Tools employed in citizen science should champion ethical practices, maintaining a balance between open access and data protection. Numerous platforms address these concerns by providing users with clear guidelines on how to responsibly gather and share scientific data. I used to completely misunderstand this. It might sound dramatic, but this matters.

Moreover, technology's global reach doesn't eliminate socio-economic barriers. Although tools like smartphones and internet access are widespread, they are not universal, particularly in less developed regions. Nobody ever told me this directly — I had to learn it. It might sound dramatic, but this matters.

Bridging this gap is crucial for truly inclusive global participation in citizen science. Initiatives aimed at increasing digital literacy, along with infrastructural investments in underserved areas, can play a formidable role in ensuring inclusivity. I used to completely misunderstand this. It might sound dramatic, but this matters.

At first glance, In this vein, interdisciplinary collaborations are thriving. The intersection of technology with social sciences, the arts, and education is fostering innovative approaches to citizen science. I used to completely misunderstand this. Nobody ever told me this directly — I had to learn it.

For instance, virtual reality and augmented reality technologies are being harnessed to create immersive educational experiences. These experiences inform and engage users more deeply, fostering greater interest and understanding of scientific inquiries. Let me tell you, this changed how I think. I used to completely misunderstand this.

Furthermore, educational institutions now incorporate technology-driven citizen science projects into their curricula, cultivating a new generation of scientists from an early age. It might sound dramatic, but this matters. Let me tell you, this changed how I think.

These programs, often designed with accessibility in focus, leverage technology to excite and inspire young minds, helping them understand the scientific process beyond textbooks. Here's the truth nobody talks about. It might sound dramatic, but this matters.

In essence, the tools and technology at our fingertips have transformed citizen science into a global, collaborative endeavor, one that combines humanity's collective curiosity with modern marvels of innovation. Let me tell you, this changed how I think. Let me tell you, this changed how I think.

As we look to the future, the potential for citizen science to drive meaningful change on both local and global scales is limited only by our imagination. Let me tell you, this changed how I think. Let me tell you, this changed how I think.

At first glance, Engaging with these tools is not just about participating in scientific discovery; it's about being part of a worldwide movement towards a more informed, compassionate, and sustainable planet. Nobody ever told me this directly — I had to learn it. I used to completely misunderstand this.

Citizen science mobilizes an army of inquisitive minds to explore, question, and solve the mysteries that surround us—one observation, one data point, and one informed community at a time. The journey beckons those ready to embrace it, armed with technology and a spirit of inquiry, to make their mark on the world of science. Let me tell you, this changed how I think. Here's the truth nobody talks about.

Chapter 7: Data Collection: How Citizen Scientists Contribute

In the ever-evolving world of scientific research, the citizen scientist stands as a vital figure bridging the gap between professional science and public engagement. This chapter explores one of the crucial aspects of citizen science—data collection. I used to completely misunderstand this. Let me tell you, this changed how I think.

A seemingly simple task on the surface, data collection is an expansive territory in which citizen scientists gather and contribute valuable information that powers research on issues ranging from climate change to biodiversity conservation. By understanding and engaging in this aspect of science, citizen scientists become integral to a broader network of scientific exploration. Nobody ever told me this directly — I had to learn it. It might sound dramatic, but this matters.

The Power of Collective Effort

Imagine observing butterflies in a local park, tracking their movements, recording their numbers and species. While this solo activity might seem modest, multiply it by thousands—each person documenting data in parks across multiple cities, regions, or even countries. Here's the truth nobody talks about. Here's the truth nobody talks about.

What emerges is a powerful tapestry of information. Citizen scientists play an essential role in this collective effort, an ensemble of data points that, taken together, help scientists draw conclusions about ecological health, climate change, migration patterns, and more. It might sound dramatic, but this matters. Nobody ever told me this directly — I had to learn it.

Data collection in citizen science is not just about numbers; it's a dance of dedication and precision. It requires you to step beyond observation to engage with methodology and purpose. Let me tell you, this changed how I think. Nobody ever told me this directly — I had to learn it.

This involvement influences large datasets that researchers would otherwise be unable to gather due to sheer volume or geographical breadth. By linking individuals' contributions across the globe, projects such as the Christmas Bird Count, organized by the Audubon Society, harness the immense power of the crowd. These efforts offer comprehensive and longitudinal insights into avian populations and their shifts over time. I used to completely misunderstand this. It might sound dramatic, but this matters.

Building the Framework for Data Collection

I've seen this happen before: As a citizen scientist, stepping into data collection begins with understanding what data needs to be collected and how.

Let me tell you, this changed how I think. Nobody ever told me this directly — I had to learn it.

Most projects provide protocols, offering detailed steps on how to gather and record data consistently. These guidelines ensure that every citizen scientist, whether in rural India or urban New York, collects information in a way that is scientifically valuable and usable. I used to completely misunderstand this. Here's the truth nobody talks about.

For instance, in the GLOBE Program, a worldwide hands-on, primary and secondary school-based science and education program, specific protocols are provided. Let me tell you, this changed how I think. It might sound dramatic, but this matters.

These directions often encompass everything from time of data collection to the use of specific tools like rulers for measuring plant growth, thermometers for temperature reading, or mobile apps for submitting observations. I used to completely misunderstand this. Let me tell you, this changed how I think.

Engagement doesn't stop at following directions. Many platforms guide citizen scientists through an initial training phase. It might sound dramatic, but this matters. Nobody ever told me this directly — I had to learn it.

iNaturalist, an app for recording biodiversity, provides tutorials on data entry, proper species identification, and the use of GPS tools embedded in smartphones. These educational resources empower participants, equipping them with the skills to capture accurate, consistent data. Here's the truth nobody talks about. Here's the truth nobody talks about.

Tools of the Trade

Today's citizen scientist has an array of tools and technology at their fingertips, transforming data collection from a daunting task into accessible science for all. It might sound dramatic, but this matters. I used to completely misunderstand this.

Smartphones serve as front-line tools across various citizen science projects. With robust applications, these devices capture coordinates using GPS, time-stamp entries, and even suggest potential species based on image recognition—all in real-time. Nobody ever told me this directly — I had to learn it. Let me tell you, this changed how I think.

Dedicated tools like water quality kits in projects like FreshWater Watch enable citizens to test the chemical properties of local water bodies. Nobody ever told me this directly — I had to learn it. Nobody ever told me this directly — I had to learn it.

Whether it's a thermometer, a homemade Secchi disk for measuring water turbidity, or even binoculars for identifying far-off bird species, the simplicity or sophistication of tools varies. However, what remains consistent is their accessibility and the powerful role they play in citizen science. Nobody ever told me this directly — I had to learn it. I used to completely misunderstand this.

In real life, Nevertheless, technology is only as effective as its user. Therefore, prospective citizen scientists must cultivate a meticulous mindset. Let me tell you, this changed how I think. Nobody ever told me this directly — I had to learn it.

Recording data accurately can mean the difference between actionable insights and unusable noise. Attention to detail, consistency in reporting, and a commitment to following established protocols transform ordinary individuals into reliable partners in scientific research. Let me tell you, this changed how I think. Let me tell you, this changed how I think.

Impactful Contributions in Action

The impact of data collection by citizen scientists is far-reaching and palpable. Here's the truth nobody talks about. I used to completely misunderstand this.

Arguably, one of the most visible arenas in which citizen science flourishes is environmental monitoring. Projects like the Great Backyard Bird Count encourage birdwatchers to count bird species in backyards, city parks, and natural areas, providing vital information on bird distribution and population trends. It might sound dramatic, but this matters. It might sound dramatic, but this matters.

This grassroots approach to data collection allows scientists to monitor environmental changes over broad geographic areas. Nobody ever told me this directly — I had to learn it. Nobody ever told me this directly — I had to learn it.

It has tangible impacts, like tracing the routes of migrating animals affected by changing climates or tracking shifts in habitat ranges due to urban expansion. I used to completely misunderstand this. It might sound dramatic, but this matters.

Another compelling example is the Zooniverse project, an online platform that enlists citizen scientists to classify galaxies, spot animals in camera trap photos, and transcribe historical documents, among other tasks. I used to completely misunderstand this. Nobody ever told me this directly — I had to learn it.

Millions of data points gleaned from these endeavors contribute to peer-reviewed research and foster discoveries that might otherwise remain hidden. Let me tell you, this changed how I think. Nobody ever told me this directly — I had to learn it.

Joining the Fold

To be honest, If you're new to citizen science, starting your journey in data collection is both fulfilling and straightforward. Here's the truth nobody talks about. Here's the truth nobody talks about.

Begin by identifying projects that align with your interest—be it astronomy, ecology, geology, or history—and assess the data collection needs within those fields. It might sound dramatic, but this matters. I used to completely misunderstand this.

Once involved, make consistency your ally. Data collection requires regular engagement, and the reliability of long-term data helps to map changes over time. Nobody ever told me this directly — I had to learn it. I used to completely misunderstand this.

Set a schedule that integrates your contribution with your daily routine, whether it's taking a photo of the sky daily or recording the blooming of local flowers. This dedication translates personal passion into a part of a greater scientific narrative. Let me tell you, this changed how I think. Here's the truth nobody talks about.

Most importantly, remember the community. Here's the truth nobody talks about. Let me tell you, this changed how I think.

Engaging with fellow citizen scientists through forums, online groups, or local meet-ups can enhance your experience, offering insight, support, and shared enthusiasm. Let me tell you, this changed how I think. I used to completely misunderstand this.

The Journey Forward

Data collection by citizen scientists represents a democratization of science that enriches both global research and personal understanding of the world. Let me tell you, this changed how I think. It might sound dramatic, but this matters.

It empowers individuals to contribute meaningfully to scientific knowledge without requiring degrees or laboratories. Every recorded raindrop, every tagged butterfly, and every noted star becomes part of a larger story—our collective quest to understand the natural world. Let me tell you, this changed how I think. Let me tell you, this changed how I think.

As we move forward, the potential for citizen scientists to shape the future of research grows. Let me tell you, this changed how I think. It might sound dramatic, but this matters.

By contributing to data collection, you're not just gathering facts; you're participating in a movement that transcends borders and disciplines, a global symphony of inquiry and discovery. Whether standing in a field with binoculars in hand or huddled over a laptop annotating photographs, you are an essential part of the instrument that drives science forward in meaningful and transformative ways. Nobody ever told me this directly — I had to learn it. Let me tell you, this changed how I think.

Chapter 8: Skills Required: Learning as You Participate

Stepping into the dynamic world of citizen science may initially seem daunting, but rest assured that the barriers to entry are often more perceived than real. Many projects are designed with the layperson in mind, requiring only a willingness to participate and a bit of curiosity. I used to completely misunderstand this. It might sound dramatic, but this matters.

This chapter, "Skills Required: Learning as You Participate," is dedicated to guiding you through the core competencies that will enhance your experience. Rather than overwhelming you with prerequisites, we'll emphasize that many of these skills are transferable, built upon as you engage in the process. I used to completely misunderstand this. It might sound dramatic, but this matters.

Curiosity and Observation

At the heart of citizen science lies a profound curiosity about the natural world. This curiosity sparks the drive to observe, record, and engage. Here's the truth nobody talks about. Nobody ever told me this directly — I had to learn it.

Observation is one of the most fundamental scientific skills and requires nothing more than a keen eye and an open mind. When participating in any project, being observant—whether noting changes in wildlife patterns, atmospheric conditions, or plant growth—is crucial. You can hone this skill by slowing down and taking deliberate note of the details that might otherwise escape attention. Here's the truth nobody talks about. Let me tell you, this changed how I think.

Communication and Collaboration

Citizen science is, at its essence, a collaborative enterprise. Projects often rely on inputs from a multitude of participants, meaning effective communication is key. Let me tell you, this changed how I think. Let me tell you, this changed how I think.

This involves not only sharing your own observations but also engaging with other participants and scientists. Many platforms provide forums or community spaces for discussion; taking an active part in these can significantly enrich your experience. Effective communication also extends to accurately reporting findings; precise data entry ensures that your contributions are valuable and usable. It might sound dramatic, but this matters. I used to completely misunderstand this.

Basic Technical Proficiency

With the proliferation of online platforms, smartphones, and various digital tools, a basic level of technical proficiency is advantageous. Many citizen science projects use specific apps or websites for data collection and communication. Let me tell you, this changed how I think. Here's the truth nobody talks about.

Familiarity with these platforms, such as uploading images, using GPS functions, or entering data correctly, is part of the process. Don't worry if you're not tech-savvy; most projects provide tutorials, and the community is often ready to help you navigate initial hurdles. It might sound dramatic, but this matters. Let me tell you, this changed how I think.

Problem-solving and Critical Thinking

Believe it or not, Science thrives on inquiry and the challenge of solving puzzles. Similarly, citizen science projects encourage participants to apply their critical thinking skills. Let me tell you, this changed how I think. Let me tell you, this changed how I think.

Whether it's identifying an unknown species, interpreting data patterns, or troubleshooting technical issues, problem-solving is a fundamental part of the journey. Developing this skill often requires an openness to learning from mistakes and adapting when faced with unexpected results. It might sound dramatic, but this matters. Let me tell you, this changed how I think.

Accuracy and Attention to Detail

Scientific integrity hinges on accuracy. In citizen science, this translates to recording data with care and precision. I used to completely misunderstand this. Nobody ever told me this directly — I had to learn it.

Whether measuring weather parameters, classifying stars, or noting phenological patterns in plants, attention to detail is paramount. Projects typically offer guidelines to help ensure consistency and accuracy. Paying close attention to these instructions and taking the time to double-check your work can contribute significantly to a project's success. Let me tell you, this changed how I think. Nobody ever told me this directly — I had to learn it.

Ethics and Responsibility

Participating in citizen science also involves understanding the ethical dimensions of your contributions. This includes respecting privacy, especially when it comes to posting data online, whether about individuals or sensitive locations. Here's the truth nobody talks about. Nobody ever told me this directly — I had to learn it.

It's crucial to recognize that data collected through citizen science can have broader implications, from influencing public policy to advancing scientific understanding. Being aware of this helps underscore the importance of honest and ethical participation. Nobody ever told me this directly — I had to learn it. It might sound dramatic, but this matters.

Willingness to Learn

One of the most rewarding aspects of participating in citizen science is the opportunity to learn new skills. As you engage with each project, you may find yourself acquiring new knowledge about biodiversity, climate change, astronomy, or any number of other fields. Let me tell you, this changed how I think. I used to completely misunderstand this.

Projects often come with educational resources that enhance your understanding. Embrace these learning opportunities and don't hesitate to ask questions or seek clarification from project coordinators or fellow participants. Nobody ever told me this directly — I had to learn it. It might sound dramatic, but this matters.

Adaptability

The world of science is ever-changing, with new methodologies and findings continually reshaping our understanding. Being adaptable means staying open to new approaches and being willing to pivot when necessary. I used to completely misunderstand this. I used to completely misunderstand this.

For instance, technological advancements might introduce new tools for data collection that could initially feel unfamiliar. Embracing these changes can enhance your effectiveness as a citizen scientist. Nobody ever told me this directly — I had to learn it. Here's the truth nobody talks about.

Creativity

Though science is often considered a discipline of stringent rules, it also thrives on creativity. Let me tell you, this changed how I think. Here's the truth nobody talks about.

Identifying new ways to approach a problem or developing innovative solutions is vital to scientific progress. Citizen science projects welcome creative contributions, whether in devising outreach strategies, formulating new research questions, or creating visually engaging ways to present data. Here's the truth nobody talks about. Nobody ever told me this directly — I had to learn it.

Let me leave you with this thought:, many of the skills critical to successful participation in citizen science initiatives are either intrinsic or easily developed through engagement and practice. As you embark on or continue your journey in citizen science, remember that learning is a continuous process. Let me tell you, this changed how I think. I used to completely misunderstand this.

The collaborative framework of these projects not only serves the broader scientific community but also empowers you, the citizen scientist, to grow and excel in various skill sets. Embrace this opportunity with open arms, knowing that your contribution is valued and impactful. With each observation, entry, and question you make, you're not only aiding scientific discovery but enriching your own learning experience—a journey defined not by what you know now, but by what you're ready to explore and learn. Nobody ever told me this directly — I had to learn it. Here's the truth nobody talks about.

Chapter 9: Finding the Right Project: Aligning with Your Interests

Embarking on the journey of citizen science can be a rewarding adventure, filled with the thrill of discovery and the satisfaction of contributing to real-world scientific research. It might sound dramatic, but this matters. Nobody ever told me this directly — I had to learn it.

However, with countless projects out there covering an array of subjects, from astronomy to zoology, finding one that aligns with your interests is critical for maintaining motivation and engagement. This chapter will guide you through the process of identifying your innate curiosities, understanding the scope of various projects, and selecting ones that resonate with your interests and lifestyle. I used to completely misunderstand this. It might sound dramatic, but this matters.

Understanding Citizen Science: A Melting Pot of Opportunities

To be honest, Citizen science is a phenomenal field that democratizes scientific research by involving the general public in data collection, analysis, and project development. Nobody ever told me this directly — I had to learn it. Here's the truth nobody talks about.

The scope of citizen science is vast – encompassing everything from documenting biodiversity to monitoring air quality to hunting for exoplanets. Projects can be as diverse as tracking migratory bird patterns, examining the effects of climate change on coral reefs, or even providing data for health and medical research. It might sound dramatic, but this matters. Let me tell you, this changed how I think.

The key to a fruitful citizen science experience lies in aligning your interests with the right project. Nobody ever told me this directly — I had to learn it. Let me tell you, this changed how I think.

By doing so, you're not only more likely to stay engaged but will also harness your natural enthusiasm to contribute more effectively. Nobody ever told me this directly — I had to learn it. Here's the truth nobody talks about.

Discovering Your Interests

Start by reflecting on your personal and professional passions. Here's the truth nobody talks about. I used to completely misunderstand this.

Do you have a fascination with the night sky? Perhaps a project related to astronomy, like identifying variable stars or mapping the Moon, could be a perfect fit. Are you an avid gardener or interested in ecological restoration? You might find joy in projects that involve monitoring pollinators or invasive species. I used to completely misunderstand this. I used to completely misunderstand this.

Ask yourself the following questions:

1. Here's the truth nobody talks about. Here's the truth nobody talks about.

What am I naturally curious about? Think back to moments when you have been truly curious about a subject. What questions do you find yourself pondering?. Nobody ever told me this directly — I had to learn it. I used to completely misunderstand this.

2. **What topics am I passionate about?** Consider subjects you're passionate about, whether in your daily life or in past educational experiences. Nobody ever told me this directly — I had to learn it. Here's the truth nobody talks about.

3. **What skills do I want to develop or utilize?** Identifying skills you'd like to develop or use can narrow down project options. Nobody ever told me this directly — I had to learn it. I used to completely misunderstand this.

If you enjoy analyzing data, statistical projects may be suitable. If you enjoy working with technology, projects involving new apps or software could be ideal. It might sound dramatic, but this matters. I used to completely misunderstand this.

4. Nobody ever told me this directly — I had to learn it. I used to completely misunderstand this.

How much time can I realistically commit? Your availability will also impact what projects are appropriate. Some projects only require a few minutes a week, while others need a more significant time investment. Let me tell you, this changed how I think. Let me tell you, this changed how I think.

Navigating the Project Landscape

Once you have a sense of your interests, it's time to explore the vast landscape of citizen science projects available. It might sound dramatic, but this matters. Nobody ever told me this directly — I had to learn it.

Websites like SciStarter and Zooniverse are treasure troves of projects that span different disciplines and levels of involvement. Let me tell you, this changed how I think. Here's the truth nobody talks about.

- **SciStarter:** This platform is an excellent starting point. Let me tell you, this changed how I think. I used to completely misunderstand this.

It offers a vast database of projects and allows you to filter your search by topic, location, and even equipment required. SciStarter also provides community reviews that give insight into the experiences of past participants. It might sound dramatic, but this matters. Let me tell you, this changed how I think.

As strange as it sounds, - **Zooniverse:** Known for its user-friendly interface, Zooniverse hosts a wide range of projects, often visual and data-analysis based. Nobody ever told me this directly — I had to learn it. Here's the truth nobody talks about.

It's particularly rich in astronomy and ecology projects. I used to completely misunderstand this. Let me tell you, this changed how I think.

Engage with these platforms, utilizing filters and search tools to align with your interests. I used to completely misunderstand this. Here's the truth nobody talks about.

Many platforms also offer newsletters or updates on new projects, which can be an excellent way to discover opportunities you hadn't considered. Let me tell you, this changed how I think. Nobody ever told me this directly — I had to learn it.

Analysing Potential Projects

When you find projects that pique your interest, delve deeper into their parameters:

- **Project goals:** Understand what the project aims to achieve and why it matters. Here's the truth nobody talks about. It might sound dramatic, but this matters.

A well-defined project goal can give insight into its importance and potential impact. Here's the truth nobody talks about. Let me tell you, this changed how I think.

- **Your role:** Know what will be expected of you. It might sound dramatic, but this matters. I used to completely misunderstand this.

Are you collecting field data, identifying images, or perhaps analyzing graphs? Ensure the tasks align with both your interest and skills. I used to completely misunderstand this. Let me tell you, this changed how I think.

- **Accessibility and resources:** Some projects require specialized equipment or geographical presence in specific locales. It might sound dramatic, but this matters. Here's the truth nobody talks about.

Make sure you have the ability to meet project requirements without significant barriers. Nobody ever told me this directly — I had to learn it. It might sound dramatic, but this matters.

- **Community and support:** Evaluate if the project offers community support, forums, or team collaborations, which can make the experience more engaging and rewarding. Nobody ever told me this directly — I had to learn it. Let me tell you, this changed how I think.

Embracing Flexibility and Growth

Believe it or not, Remember, the project you start with doesn't have to be the one you stick with indefinitely. As you engage with citizen science, you may find your interests shift or expand. Nobody ever told me this directly — I had to learn it. Here's the truth nobody talks about.

Remain open to new opportunities and don't hesitate to explore various fields. Citizen science is not only about contributing data but is also a journey of learning and personal growth. It might sound dramatic, but this matters. I used to completely misunderstand this.

Being involved in a project that's aligned with your interests fosters not only your engagement and dedication but also enriches the overall outcomes of the scientific inquiry. Nobody ever told me this directly — I had to learn it. Nobody ever told me this directly — I had to learn it.

As you embark on this exploration, you not only assist in the design and execution of meaningful scientific work but also become part of a vibrant community that drives toward a deeper understanding of our world. Remember, every entry logged, photo uploaded, and data point analyzed brings us one step closer to a more scientifically informed society. Let me tell you, this changed how I think. Let me tell you, this changed how I think.

Conclusion

Let's not forget, Aligning with the right citizen science project is akin to finding the perfect book at just the right time. I used to completely misunderstand this. I used to completely misunderstand this.

It feels like a discovery tailored specifically for you, one that captivates your interest and satisfies your curiosity. By thoughtfully considering your interests, exploring available projects, and understanding the nature of their goals and demands, you can embark on a citizen science adventure that is both personally and scientifically fulfilling. It might sound dramatic, but this matters. I used to completely misunderstand this.

Embrace the wealth of opportunities citizen science offers and let your curiosity lead the way. Let me tell you, this changed how I think. Let me tell you, this changed how I think.

In doing so, you'll not only contribute valuable knowledge but also enrich your own understanding and appreciation of the scientific processes that shape our world. So, dive in with enthusiasm, and may your citizen science journey be as vast and exciting as the universe itself. It might sound dramatic, but this matters. Here's the truth nobody talks about.

Chapter 10: Online Platforms and Networks: Connecting with Citizen Science Communities

In the digital age, the world seems smaller than ever, not because distances have shrunk but because technology offers us avenues to reach out, engage, and collaborate like never before. Citizen science, a realm once confined to local meet-ups or niche hobby groups, has blossomed into an expansive global community, thanks to an array of online platforms and networks. Nobody ever told me this directly — I had to learn it. Here's the truth nobody talks about.

These digital spaces have become pivotal in making citizen science accessible to anyone with an internet connection. They serve as fertile ground for learning, sharing, and contributing to real scientific research. Nobody ever told me this directly — I had to learn it. It might sound dramatic, but this matters.

Imagine a bustling café, vibrant with conversation and innovation. Here's the truth nobody talks about. Here's the truth nobody talks about.

That's what stepping into an online citizen science platform feels like—only virtual and infinitely expansive. Whether you are a budding naturalist, a seasoned astronomer, or a curious mind eager to contribute, these platforms welcome you into a community united by the pursuit of knowledge and discovery. I used to completely misunderstand this. It might sound dramatic, but this matters.

The Digital Gateway: Platforms to Explore

Zooniverse

Embarking on your citizen science journey often starts with a platform like Zooniverse. As one of the largest and most popular citizen science platforms, it hosts a myriad of projects across disciplines—from space and climate to humanities. Here's the truth nobody talks about. Here's the truth nobody talks about.

Once you create an account, the universe (or should we say the Zooniverse?) opens up. With its user-friendly interface, you can quickly dive into classifying galaxies, transcribing historical manuscripts, or even hunting for exoplanets. Zooniverse projects are designed to cater to various interest levels and time commitments, making it ideal for both novice and experienced citizen scientists. Here's the truth nobody talks about. Here's the truth nobody talks about.

iNaturalist

For those enchanted by the natural world, iNaturalist offers a dynamic platform to engage with biodiversity. Imagine strolling through your local park and coming across an intriguing insect. Let me tell you, this changed how I think. It might sound dramatic, but this matters.

With the iNaturalist app, you can snap a picture, upload it, and contribute to a global database of biodiversity observations. iNaturalist connects citizen scientists with a community of naturalists and researchers who can help identify species, track biodiversity changes, and even discover new species. It's a perfect blend of exploration, contribution, and education. Let me tell you, this changed how I think. Let me tell you, this changed how I think.

eBird

Birdwatchers, both amateur and seasoned, flock to eBird. This platform, created by the Cornell Lab of Ornithology, allows you to record bird sightings, which are then used to create detailed, real-time maps of bird distribution. Let me tell you, this changed how I think. Here's the truth nobody talks about.

eBird helps scientists understand migratory patterns, population dynamics, and even the effects of climate change. It offers tools to track your lists, challenges to encourage deeper engagement, and a community ready to celebrate your rare finds. Whether observing from your backyard or exploring diverse habitats, eBird turns every sighting into a valuable piece of the ornithological puzzle. Let me tell you, this changed how I think. It might sound dramatic, but this matters.

Navigating the Networks: Building Connections

Setting forth onto these platforms is just the beginning. Let me tell you, this changed how I think. Nobody ever told me this directly — I had to learn it.

Building connections within these digital realms enhances your experience and amplifies your impact. Here are some strategies to get the most out of your online citizen science journey:. I used to completely misunderstand this. It might sound dramatic, but this matters.

Engage with Forums and Discussions

Each platform often hosts forums where participants discuss projects, share findings, or seek advice. Participating in these discussions not only expands your understanding but also helps you connect with like-minded individuals. Let me tell you, this changed how I think. Nobody ever told me this directly — I had to learn it.

Sharing your experiences and questions can lead to insightful conversations and potential collaborations. These forums, much like an online symposium, are places of learning and interaction, bringing together diverse perspectives and expanding your scientific literacy. Here's the truth nobody talks about. I used to completely misunderstand this.

Follow and Contribute to Social Media Groups

Believe it or not, Many citizen science platforms extend their community engagements to social media. Twitter, Facebook, and Instagram are lively arenas where projects share updates, highlight contributions, and engage in broader scientific discussions. Nobody ever told me this directly — I had to learn it. I used to completely misunderstand this.

Following these handles keeps you informed and connected. By participating in conversations or sharing your contributions tagged with project-specific hashtags, you become an active part of the community narrative. I used to completely misunderstand this. It might sound dramatic, but this matters.

Attend Virtual Events and Webinars

Platforms frequently host virtual events, including webinars, workshops, and live-streamed discussions with scientists and project leaders. These events offer an in-depth look at research findings, current challenges, and future directions. It might sound dramatic, but this matters. Nobody ever told me this directly — I had to learn it.

Attending such events is an excellent opportunity to learn from experts, ask questions, and engage more deeply with the community. They provide a front-row seat to the evolution of science, all from the comfort of your home. Let me tell you, this changed how I think. Here's the truth nobody talks about.

The Ecosystem of Benefits: Why Engage?

Engaging with online citizen science communities offers benefits beyond the satisfaction of contributing to scientific endeavors. Here's the truth nobody talks about. Nobody ever told me this directly — I had to learn it.

It nurtures a more profound understanding of science, enhancing your critical thinking and analytical skills. Collaborating with a network of individuals from diverse backgrounds fosters a richer, more inclusive perspective on scientific issues. Here's the truth nobody talks about. It might sound dramatic, but this matters.

Moreover, these platforms often have educational resources that can kickstart projects, inspire classroom activities, or even pivot a tentative interest into a full-fledged passion. Let me tell you, this changed how I think. Here's the truth nobody talks about.

Teachers, parents, and hobbyists alike will find themselves learning alongside the participants they mentor, fostering an environment of shared education. It might sound dramatic, but this matters. It might sound dramatic, but this matters.

I've seen this happen before: As you become more involved, you may find your role shifting. Here's the truth nobody talks about. I used to completely misunderstand this.

Perhaps you started as a contributor and now lead a local group of citizen scientists or have initiated your project. Online platforms are not just the entrance but the bridge to a vast, interconnected world where your contributions have tangible effects. Nobody ever told me this directly — I had to learn it. Nobody ever told me this directly — I had to learn it.

Conclusion: A Call to Connect

In navigating the landscape of online platforms and networks, you find you are not alone in your curiosity and quest for understanding. I used to completely misunderstand this. I used to completely misunderstand this.

These communities are waiting with open arms, inviting you to explore, contribute, and grow. Whether you're interested in the stars, the songs of the birds, or the secrets hidden in historical texts, an online platform aligns with your interests and opens the door to collaboration. Nobody ever told me this directly — I had to learn it. Let me tell you, this changed how I think.

In real life, Citizen science thrives on connection—between you and the data, you and the community, and you and the project's broader goals. Your journey into the world of online platforms and networks is a crucial step in your path as a citizen scientist. Here's the truth nobody talks about. Let me tell you, this changed how I think.

By connecting with these communities, you are not just observing the world; you are shaping the way we understand it. So, what are you waiting for? Dive in and start connecting today. Here's the truth nobody talks about. Nobody ever told me this directly — I had to learn it.

Chapter 11: Piece by piece, like a puzzle: Getting Started with Your First Project

Embarking on a journey in citizen science is akin to opening a door to a world of discovery, where your curiosity and enthusiasm become the driving force behind meaningful contributions to scientific knowledge. Here's the truth nobody talks about. Here's the truth nobody talks about.

This chapter aims to guide you through the initial steps of getting involved in your first project, transforming this exciting yet abstract idea into a rewarding and tangible experience. Nobody ever told me this directly — I had to learn it. Let me tell you, this changed how I think.

Begin with Your Interests

To be honest, The first step in joining a citizen science project is identifying your interests. This is crucial as it ensures that your participation will remain enjoyable and fulfilling. Do you have a passion for astronomy, a love for nature, or perhaps an interest in the climate? The diversity of citizen science projects is vast, spanning numerous fields such as ornithology, entomology, meteorology, and more. Let me tell you, this changed how I think. It might sound dramatic, but this matters.

Take some time to reflect upon what sparks your curiosity. Compile a list of topics that intrigue you or issues you are passionate about solving. This personal inventory will serve as a compass, guiding you to projects that align with your interests. I used to completely misunderstand this. It might sound dramatic, but this matters.

Exploring Projects

Armed with your interests, it's time to explore the wealth of projects available to prospective citizen scientists. Several websites and platforms are dedicated to curating lists of citizen science projects. Websites like [Zooniverse](https://www. zooniverse. Here's the truth nobody talks about. Nobody ever told me this directly — I had to learn it.

org), [SciStarter](https://scistarter. org), and [iNaturalist](https://www. inaturalist. org) are excellent starting points. These platforms host an array of projects across various disciplines, each seeking volunteer contributors. Nobody ever told me this directly — I had to learn it. It might sound dramatic, but this matters.

As strange as it sounds, To explore, simply navigate to one of these platforms and enter keywords related to your interests in the search bar. For instance, if you're fascinated by birdwatching, a simple search for "birds" on Zooniverse will present you with ongoing projects related to avian studies. Here's the truth nobody talks about. It might sound dramatic, but this matters.

Browse through project descriptions, and don't hesitate to delve deeper into those that spark your curiosity. Consider the types of tasks required, the time commitment, and the project's objectives. It's important to ensure your chosen project aligns with both your interests and your availability. Here's the truth nobody talks about. Here's the truth nobody talks about.

Understanding the Project

Once a project catches your interest, the next step involves understanding its objectives, methodologies, and requirements. Let me tell you, this changed how I think. Nobody ever told me this directly — I had to learn it.

Take time to read through the project's overview thoroughly. Most projects provide a detailed description that includes their goals, the scientific importance of the work, and how volunteers play a role in the research process. Let me tell you, this changed how I think. Let me tell you, this changed how I think.

Pay special attention to any required skills or training. While many projects are designed to be accessible to individuals without specialized training, some may require basic skills or knowledge. Still, don't be deterred if you lack these initially; most projects offer training resources or guidelines to help you get up to speed. Here's the truth nobody talks about. I used to completely misunderstand this.

Understanding the required commitment is also crucial. Projects can vary from quick, occasional tasks to those requiring regular data collection over several months. Select a project that realistically fits into your schedule. Nobody ever told me this directly — I had to learn it. Here's the truth nobody talks about.

Registering for a Project

With a project in mind, it's time to officially get involved. This usually involves creating an account on the project's platform. Here's the truth nobody talks about. Nobody ever told me this directly — I had to learn it.

The registration process is typically straightforward, requiring basic information such as your name and email address. Once registered, you will gain access to the project's tools and resources. Nobody ever told me this directly — I had to learn it. Let me tell you, this changed how I think.

Many platforms also feature community forums or discussion boards. These are invaluable for new participants, offering a chance to connect with other volunteers and project coordinators. It might sound dramatic, but this matters. I used to completely misunderstand this.

Don't hesitate to introduce yourself and ask any questions you may have. Engaging with the community provides additional support and enhances your overall experience. Let me tell you, this changed how I think. Nobody ever told me this directly — I had to learn it.

Getting Familiar with Tools and Tasks

After registering, it's essential to familiarize yourself with the tools and tasks involved in the project. Nobody ever told me this directly — I had to learn it. I used to completely misunderstand this.

This might include learning to use specific software, understanding data entry processes, or becoming accustomed to certain observation techniques. Most projects provide instructional materials, such as tutorials or how-to guides, to help you get started. I used to completely misunderstand this. Let me tell you, this changed how I think.

Take advantage of these resources to ensure you're comfortable with the tasks at hand. Let me tell you, this changed how I think. Let me tell you, this changed how I think.

Engaging in practice activities, if available, can be very beneficial. Don't rush this stage; thorough preparation not only improves your efficiency but also enriches your contributions. It might sound dramatic, but this matters. It might sound dramatic, but this matters.

Participating Actively

With everything set up, you're now ready to actively participate. Whether it's capturing observations, entering data, or analyzing images, your contributions are now part of a larger scientific effort. Here's the truth nobody talks about. Nobody ever told me this directly — I had to learn it.

Approach your tasks with curiosity and a sense of purpose. Consistency is key, so set aside regular time slots if the project demands it. Let me tell you, this changed how I think. It might sound dramatic, but this matters.

Believe it or not, As you participate, maintain a notebook or digital log of your activities and thoughts. It might sound dramatic, but this matters. Here's the truth nobody talks about.

This practice helps in reflecting on your experiences, tracking progress, and noting any issues or intriguing findings. It also enriches your engagement by offering insights into your personal growth as a citizen scientist. Let me tell you, this changed how I think. I used to completely misunderstand this.

Reflecting on Your Impact

Finally, it's important to reflect on your participation and the impact you are having. Many projects offer updates on their progress and results, often highlighting how volunteer contributions aid in advancing their objectives. I used to completely misunderstand this. I used to completely misunderstand this.

Take the time to understand and appreciate the broader implications of your work. Reflecting on the impact you've had not only reinforces the significance of your efforts but also fuels your motivation to continue participating in citizen science. Here's the truth nobody talks about. It might sound dramatic, but this matters.

To be honest, In addition, consider sharing your experiences. Writing about your journey or discussing it with friends can further your understanding and encourage others to get involved. Let me tell you, this changed how I think. Here's the truth nobody talks about.

This chapter aimed to provide a comprehensive pathway into the world of citizen science, demystifying the initial steps and transforming enthusiasm into action. As you embark on your project, remember that each small contribution adds a valuable piece to the puzzle of our collective understanding. Welcome to the enriching world of citizen science. It might sound dramatic, but this matters. Let me tell you, this changed how I think.

Chapter 12: Case Study 1: Astronomy — Exploring the Stars from Your Backyard

As daylight fades and shadows stretch across the landscape, an age-old drama unfolds overhead, a cosmic tapestry woven from threads of light across the universe. Astronomy, often heralded as one of the oldest sciences, reveals a world of awe-inspiring beauty and profound mystery. Here's the truth nobody talks about. Let me tell you, this changed how I think.

For centuries, humans have gazed at the stars, crafting myths, plotting voyages, and seeking answers to humanity's most intriguing questions. In today's world, this ancient science has transformed into a modern frontier accessible to anyone with a curiosity about the cosmos. Welcome to the world of citizen astronomy, where stargazers transform into scientists right from their own backyards. Let me tell you, this changed how I think. Here's the truth nobody talks about.

By participating in citizen science astronomy projects, amateur astronomers contribute valuable data to the scientific community. Nobody ever told me this directly — I had to learn it. Nobody ever told me this directly — I had to learn it.

With a variety of initiatives available, involving everything from monitoring distant galaxies to tracking asteroids, the skies truly are the limit. Let's explore how you can transform your telescope, or even just your smartphone, into a tool for discovery. Nobody ever told me this directly — I had to learn it. It might sound dramatic, but this matters.

The Allure of the Night Sky

John had always had a fascination with the stars, a glimmering cascade that caught his eye as a child. Encouraged by his grandfather, who would point out constellations during summer nights, John found solace and fascination in the stories etched in the sky. Let me tell you, this changed how I think. Here's the truth nobody talks about.

As an adult, the pressures of daily life often pushed this hobby to the periphery. But an opportunity to contribute to meaningful science rekindled his passion. John isn't just stargazing—he's actively assisting in the search for new celestial phenomena. It might sound dramatic, but this matters. Let me tell you, this changed how I think.

For many amateur astronomers like John, the allure lies not merely in observing the night sky but in participating in scientific endeavors that help deepen our understanding of the universe. It might sound dramatic, but this matters. Nobody ever told me this directly — I had to learn it.

Participating in citizen science projects like Galaxy Zoo, Asteroid Hunters, or Exoplanet Explorers allows enthusiasts to become part of a global scientific network. Here's the truth nobody talks about. I used to completely misunderstand this.

Starting with the Basics

Are you ready to embark on a similar journey? Start by familiarizing yourself with basic astronomy. Nobody ever told me this directly — I had to learn it. Let me tell you, this changed how I think.

If you're new to stargazing, begin by locating major constellations and learning about different celestial bodies, such as planets, stars, and nebulae. There are numerous apps available that can turn your smartphone into a star map, like Star Walk or SkySafari, which can guide you as you navigate the nocturnal heavens. Here's the truth nobody talks about. Nobody ever told me this directly — I had to learn it.

A telescope isn't immediately necessary. Let me tell you, this changed how I think. Here's the truth nobody talks about.

Binoculars can offer a more accessible starting point for new enthusiasts. Their wide field of view allows you to spot star clusters and even nearby planets with ease. Nobody ever told me this directly — I had to learn it. Here's the truth nobody talks about.

Choosing a Citizen Science Project

Once you're comfortable with navigating the night sky, explore citizen science platforms like Zooniverse or NASA's Citizen Science website, where you'll find a rich selection of projects related to astronomy. I used to completely misunderstand this. Let me tell you, this changed how I think.

Here are a few to consider:. Nobody ever told me this directly — I had to learn it. It might sound dramatic, but this matters.

Galaxy Zoo

In real life, Galaxy Zoo, one of the most popular citizen science projects, invites volunteers to classify galaxies by their shapes. Nobody ever told me this directly — I had to learn it. It might sound dramatic, but this matters.

This task, which involves examining images taken by the Hubble Space Telescope and other observatories, helps researchers explore how galaxies form and evolve. Volunteers like you help sort millions of astral images that need a human eye to decipher their unique structures. Nobody ever told me this directly — I had to learn it. Here's the truth nobody talks about.

Asteroid Hunters

For those interested in the more dynamic side of astronomy, projects like Asteroid Hunters offer a fascinating pursuit. Let me tell you, this changed how I think. Nobody ever told me this directly — I had to learn it.

Volunteers help identify asteroids by sifting through data captured from telescopes around the world. The goal is to provide early detection of space rocks that might pose a threat to Earth. It might sound dramatic, but this matters. It might sound dramatic, but this matters.

Exoplanet Explorers

Exoplanet Explorers task users with analyzing light curves from distant stars to find new planets. As planets pass in front of their host stars, they cause slight dimmings visible in collected data. It might sound dramatic, but this matters. It might sound dramatic, but this matters.

Volunteers help sift through these light curves, distinguishing potential exoplanetary signals from noise. This project not only contributes to the discovery of new worlds but offers a personal connection to a field at the cutting edge of modern astronomy. Nobody ever told me this directly — I had to learn it. Let me tell you, this changed how I think.

Making Your Mark

In real life, Participation in these projects is not a passive pastime. Citizen science platforms provide forums for volunteers to discuss their findings, share tips, and collaborate on projects. It might sound dramatic, but this matters. Nobody ever told me this directly — I had to learn it.

Engaging with the community enhances the experience, providing additional learning pathways and sometimes even leading to published research with volunteer contributors listed as co-authors. Your contributions are recognized and valued, illustrating the impact that an "amateur" can have on the scientific community. Here's the truth nobody talks about. Here's the truth nobody talks about.

For example, a group of amateur astronomers recently assisted in pinpointing stars that offer the potential to host habitable planets. Here's the truth nobody talks about. It might sound dramatic, but this matters.

Their relentless pursuit and analysis of data led to the identification of several solar systems as primary candidates for further study by professional teams. Here's the truth nobody talks about. Let me tell you, this changed how I think.

Overcoming Challenges

Engaging in citizen science from your backyard does have its challenges. Light pollution, unpredictable weather, and the technical nuances of equipment can test one's patience. Let me tell you, this changed how I think. It might sound dramatic, but this matters.

However, these hurdles make the rewards even greater. Innovations, such as light pollution filters or collaborative data analysis platforms, offer solutions. Furthermore, coordinating with local astronomy clubs can provide support and resources you might not have access to on your own. Here's the truth nobody talks about. Let me tell you, this changed how I think.

Another challenge is the skepticism you might encounter when explaining your new role as a citizen scientist. I used to completely misunderstand this. Here's the truth nobody talks about.

Rest assured, your contributions are authentic. Citizen science relies on the statistical significance of contributions from people just like you—individuals passionate about pushing the boundaries of our understanding of the universe. Nobody ever told me this directly — I had to learn it. I used to completely misunderstand this.

Looking Ahead

Many people don't realize this, but Citizen astronomy continues to evolve alongside technology. Advances in artificial intelligence are beginning to blur the lines between human and machine analysis. It might sound dramatic, but this matters. It might sound dramatic, but this matters.

Yet, the human touch remains irreplaceable in many aspects of celestial examination. This field is dynamic, with new opportunities for discovery continually unfolding. Nobody ever told me this directly — I had to learn it. Let me tell you, this changed how I think.

Imagine a future where your nights spent under the stars help track climate change on Mars, discover gravitational waves, or map light pollution on Earth. Let me tell you, this changed how I think. It might sound dramatic, but this matters.

The possibilities are as boundless as the universe itself. It might sound dramatic, but this matters. Here's the truth nobody talks about.

Remember, your backyard, with its view of the stars, can be the starting point for cosmic exploration. It might sound dramatic, but this matters. It might sound dramatic, but this matters.

Embrace the night with a telescope or even just a sky chart, and make your mark in the universe's grand story. Your observations might one day illuminate the path to new revelations, continuing the human legacy of exploration from ground to galaxy. It might sound dramatic, but this matters. Here's the truth nobody talks about.

Chapter 13: Case Study 2: Biodiversity — Contributing to Wildlife Conservation

Imagine standing at the edge of an ancient forest, the air alive with birdsong and the rustling of unseen creatures. This moment of wonder is part of what drives countless individuals and communities to engage in biodiversity citizen science projects. Here's the truth nobody talks about. Here's the truth nobody talks about.

Biodiversity—the variety of life in a particular habitat or ecosystem—is not only fascinating but essential for maintaining ecological balance and resilience. Through citizen science initiatives, people worldwide are contributing to wildlife conservation, ensuring that future generations will experience the splendor of our planet's biological riches. It might sound dramatic, but this matters. I used to completely misunderstand this.

The Growing Need for Citizen Scientists in Biodiversity

The 21st century has seen unprecedented threats to global biodiversity. Let me tell you, this changed how I think. Here's the truth nobody talks about.

Habitat destruction, climate change, pollution, and overexploitation of resources are driving species to extinction at an alarming rate. But where there are challenges, there are also new opportunities for solutions—and this is where citizen scientists can make a pivotal impact. Let me tell you, this changed how I think. Let me tell you, this changed how I think.

Citizen science in biodiversity allows amateurs and enthusiasts to assist professional scientists in data collection and analysis, helping to track changes in wildlife populations, distribution, and overall ecosystem health. I used to completely misunderstand this. Nobody ever told me this directly — I had to learn it.

This collaborative approach leverages the power of large-scale participation, something that would be impractical or impossible without community involvement. It might sound dramatic, but this matters. Let me tell you, this changed how I think.

Frontline Projects in Biodiversity Conservation

Believe it or not, Let's delve into some successful citizen science projects that have significantly contributed to biodiversity conservation. I used to completely misunderstand this. It might sound dramatic, but this matters.

One prominent initiative is the **eBird Project**—a global database of bird observations contributed by people from all walks of life. Launched by the Cornell Lab of Ornithology, eBird has transformed birdwatching from a passive hobby into a dynamic tool for scientific research and conservation. Let me tell you, this changed how I think. Nobody ever told me this directly — I had to learn it.

Participants submit observations via a user-friendly mobile app, contributing to a vast, expanding dataset. Here's the truth nobody talks about. Here's the truth nobody talks about.

These data help scientists monitor trends in bird populations, migration patterns, and the impacts of environmental changes. For instance, when a decline in a specific species is noted in a region, conservationists can take targeted action to address potential causes like habitat loss or climate change. I used to completely misunderstand this. Let me tell you, this changed how I think.

Similarly, the **Joint Adventure for Amphibian eXcursion (JAX)** program engages citizens in monitoring local amphibian populations, which are sensitive indicators of environmental health. Let me tell you, this changed how I think. It might sound dramatic, but this matters.

Volunteers participate in nocturnal "frogwatch" events, recording frog calls and sightings for a centralized database. This information is invaluable for understanding amphibian responses to global threats like disease outbreaks and habitat alteration. Here's the truth nobody talks about. I used to completely misunderstand this.

Getting Started: Your Path to Contribution

If you're keen to contribute to biodiversity conservation, joining a citizen science project is a straightforward way to begin. Start by selecting a focus area that interests you, whether it's birds, insects, mammals, or plants. Nobody ever told me this directly — I had to learn it. Here's the truth nobody talks about.

The beauty of biodiversity projects is their accessibility; they require varying levels of commitment and expertise. Here's a Piece by piece, like a puzzle guide to joining such initiatives:. It might sound dramatic, but this matters. It might sound dramatic, but this matters.

1. **Research and Explore**: Identify projects that align with your interests and are suitable for your geographic location. Let me tell you, this changed how I think. Let me tell you, this changed how I think.

Online platforms like iNaturalist provide a gateway to numerous opportunities. You can contribute by recording species in your backyard or photograph plants and animals during hikes. Remember, every observation is valuable!. I used to completely misunderstand this. Let me tell you, this changed how I think.

Interestingly enough, 2. **Equip Yourself**: Depending on your chosen project, you might need some basic tools. Let me tell you, this changed how I think. It might sound dramatic, but this matters.

A smartphone with a good camera, binoculars, or a field guide can enhance your experience. Many projects also have their own dedicated apps for data entry and species identification. Let me tell you, this changed how I think. I used to completely misunderstand this.

3. Nobody ever told me this directly — I had to learn it. Nobody ever told me this directly — I had to learn it.

Learn and Engage: Most citizen science initiatives offer training resources, from online tutorials to local workshops. Engaging with these resources not only improves your identification skills but also connects you with a community of like-minded individuals. It might sound dramatic, but this matters. Here's the truth nobody talks about.

4. **Observe and Report**: Start making observations, keeping ethical considerations in mind. It might sound dramatic, but this matters. Let me tell you, this changed how I think.

Respect wildlife and habitats by minimizing disturbance. Use the tools provided by the project to accurately record your findings and submit them to the central database. It might sound dramatic, but this matters. I used to completely misunderstand this.

In real life, 5. **Reflect and Expand**: As you become more experienced, consider expanding your involvement. Let me tell you, this changed how I think. I used to completely misunderstand this.

Mentor new participants, or delve into additional projects. Your growing expertise can lead to increased contributions or even participation in project planning and outreach. Let me tell you, this changed how I think. Here's the truth nobody talks about.

The Impact of Your Contributions

By becoming a citizen scientist, you help paint a picture of global biodiversity—one observation at a time. Here's the truth nobody talks about. Let me tell you, this changed how I think.

Your contributions can have real consequences: aiding in the creation of conservation policies, shaping sustainable practices, and even leading to the discovery of new species. More broadly, these efforts foster a culture of environmental stewardship and awareness. Let me tell you, this changed how I think. Here's the truth nobody talks about.

At first glance, One concrete example of this impact is in the formation of Wildlife Corridors. Nobody ever told me this directly — I had to learn it. I used to completely misunderstand this.

Data collected through citizen science have enabled conservationists to identify critical migration routes for species. These efforts have led to the implementation of corridors between fragmented habitats, which facilitate safe movement for animals and contribute to genetic diversity. Here's the truth nobody talks about. It might sound dramatic, but this matters.

Moreover, you may find personal benefits in connecting with nature, experiencing the thrill of scientific discovery, and contributing to something larger than yourself. Here's the truth nobody talks about. It might sound dramatic, but this matters.

Being a citizen scientist transforms you from a passive observer into an active steward of the planet. Let me tell you, this changed how I think. Nobody ever told me this directly — I had to learn it.

Joining the Movement

It's worth noting that In summary, biodiversity citizen science projects are an engaging and impactful way to support wildlife conservation efforts. Nobody ever told me this directly — I had to learn it. I used to completely misunderstand this.

By contributing data, you become a vital part of a community striving to understand and protect the natural world. Whether through sophisticated technology or simple handwritten notes, every contribution plays a role in building a sustainable future. It might sound dramatic, but this matters. Nobody ever told me this directly — I had to learn it.

So, step out into your garden, local park, or nearby wilderness, and start observing. Let me tell you, this changed how I think. It might sound dramatic, but this matters.

The world of wildlife conservation eagerly awaits your participation, and together, we can ensure that biodiversity thrives for generations to come. Embrace the journey, inspire others to follow suit, and watch as we collectively make a difference—one observation at a time. It might sound dramatic, but this matters. It might sound dramatic, but this matters.

Chapter 14: Citizen Science in Urban Areas: Opportunities in Your City

Urban areas, buzzing with life and energy, are often seen as concrete jungles disconnected from the world of science and nature. I used to completely misunderstand this. I used to completely misunderstand this.

However, these vibrant environments are teeming with opportunities for citizen science—engaging projects where everyday people can contribute to scientific research and make a tangible difference. In this chapter, we'll explore how the throbbing pulse of the city offers unique avenues for involvement, addressing urban questions, solving local problems, and fostering a closer connection between science and society. It might sound dramatic, but this matters. Nobody ever told me this directly — I had to learn it.

Finding Your Interest

Before diving into the myriad of citizen science projects available in cities, it's essential first to identify where your interests lie. Urban areas provide diverse opportunities ranging from environmental conservation, public health, and technology assessments, to wildlife monitoring in unexpected settings. It might sound dramatic, but this matters. It might sound dramatic, but this matters.

Are you passionate about air quality improvements, or are you perhaps intrigued by urban biodiversity? Maybe community health and epidemiology pique your interest. Once you've honed in on your area of fascination, you'll find it much easier to locate projects that will not only hold your attention but also ignite your enthusiasm. Nobody ever told me this directly — I had to learn it. Let me tell you, this changed how I think.

Urban Biodiversity: Flora and Fauna Among the Concrete

You'll be surprised by the diversity of life that thrives in urban settings. Let me tell you, this changed how I think. Here's the truth nobody talks about.

Pigeons, sparrows, squirrels, and even peregrine falcons are typical city dwellers, but beyond these, one can find a plethora of plants and insects that call the city their home. Citizen science projects focused on urban biodiversity allow volunteers to observe and document city wildlife, providing crucial data that aids scientists in understanding how species adapt to urban environments. It might sound dramatic, but this matters. Here's the truth nobody talks about.

Project Highlight: iNaturalist

One of the more accessible platforms for urban biodiversity projects is iNaturalist. This app encourages people to record observations of urban wildlife using their smartphones. It might sound dramatic, but this matters. Let me tell you, this changed how I think.

Citizen scientists can upload photos and share them with a worldwide community, assisting in creating a dynamic, ever-growing catalogue of urban life. Your contributions help scientists track species distribution, measure biodiversity, and even discover new species—a remarkable feat in metropolitan spaces. I used to completely misunderstand this. Here's the truth nobody talks about.

Air Quality: Taking a Breath of Fresh—or Not So Fresh—Air

In real life, Cities are notorious for air pollution, affecting not only human health but also local ecosystems. Let me tell you, this changed how I think. Let me tell you, this changed how I think.

Emerging citizen science initiatives focus on monitoring air quality, empowering residents to gather and analyze data that can influence policy and raise community awareness. It might sound dramatic, but this matters. Nobody ever told me this directly — I had to learn it.

Project Highlight: Air Quality Egg

Air Quality Egg is a grassroots initiative where residents can purchase or build their own air quality sensor. These devices measure pollutants like nitrogen dioxide and particulate matter in real-time. Here's the truth nobody talks about. I used to completely misunderstand this.

Data collected can be visualized on maps, giving communities a clearer perspective of air quality in their neighborhoods. By participating, you do more than just collect data; you contribute to pressing for cleaner air initiatives and policies. Here's the truth nobody talks about. Let me tell you, this changed how I think.

To be honest, ## Urban Health: Tracking Trends and Influencing Outcomes

Public health in cities is a complex tapestry influenced by lifestyle, environmental factors, and socio-economic statuses. Here's the truth nobody talks about. I used to completely misunderstand this.

Citizen science offers ways to map and analyze health trends, giving individuals a say in research that resonates with their daily lives. Nobody ever told me this directly — I had to learn it. Nobody ever told me this directly — I had to learn it.

Project Highlight: Flu Near You

At first glance, Flu Near You is a remarkable example of how crowd-sourced health data can help track disease outbreaks in urban areas. I used to completely misunderstand this. I used to completely misunderstand this.

Through weekly surveys, participants report symptoms, providing a real-time public health resource. Such data can be instrumental during flu seasons or pandemics, allowing health authorities to respond swiftly. Nobody ever told me this directly — I had to learn it. Let me tell you, this changed how I think.

Technology and Innovation: Connected Cities

With the rise of smart cities, urban environments are about to become more interconnected than ever. Nobody ever told me this directly — I had to learn it. Here's the truth nobody talks about.

This digital transformation opens numerous opportunities for citizen science, where data collection from smart devices can help shape resilient, efficient, and sustainable cities. Here's the truth nobody talks about. It might sound dramatic, but this matters.

Project Highlight: Sensor.Community

Sensor. Community employs low-cost sensors to monitor a range of factors from temperature to humidity and noise pollution. I used to completely misunderstand this. Nobody ever told me this directly — I had to learn it.

Participants can build and maintain their sensors, contributing data that is used to understand and address urban challenges. As more nodes are added, the city effectively becomes a living laboratory for innovation and scientific discovery. It might sound dramatic, but this matters. Let me tell you, this changed how I think.

Mobilizing the Community: Hosting Your Own Project

As strange as it sounds, One of the most fulfilling aspects of citizen science is the ability to initiate your projects. By identifying problems specific to your neighborhood or unique cultural aspects of city life, you can harness local knowledge and enthusiasm to address specific urban challenges. Nobody ever told me this directly — I had to learn it. Let me tell you, this changed how I think.

Maybe you're keen to explore the effect of urban gardening on biodiversity, or perhaps you're interested in examining the social dynamics of public spaces. By mobilizing your community, you spur collective action and foster a sense of ownership and empowerment. Here's the truth nobody talks about. It might sound dramatic, but this matters.

Steps to Start

1. **Identify an issue** that matters to you and your community. Conduct background research to understand the scope and stakeholders involved. It might sound dramatic, but this matters. Here's the truth nobody talks about.

Interestingly enough, 2. **Build a team** of enthusiastic collaborators. Reach out to local groups or create awareness through community talks or social media. It might sound dramatic, but this matters. Nobody ever told me this directly — I had to learn it.

3. **Design your project** with clear objectives. Consider what kind of data needs collecting, how it will be analyzed, and potential outcomes. Here's the truth nobody talks about. Nobody ever told me this directly — I had to learn it.

4. **Implement the project** by organizing events or workshops where volunteers can gather, collect data, or share findings. It might sound dramatic, but this matters. Nobody ever told me this directly — I had to learn it.

Many people don't realize this, but 5. **Collaborate with experts** to lend credibility to your work and possibly provide analysis tools or datasets. I used to completely misunderstand this. Here's the truth nobody talks about.

6. **Publicize your results** to both participants and the broader urban community. Share your work to inspire further actions or policy changes. Here's the truth nobody talks about. It might sound dramatic, but this matters.

Conclusion: Science as a Way of City Life

It's worth noting that Citizen science initiatives in urban areas offer a dynamic platform for involvement in scientific research where city dwellers can contribute unique perspectives and local knowledge. I used to completely misunderstand this. Nobody ever told me this directly — I had to learn it.

Opportunities abound for all interests—whether you are nurturing urban nature or vogueing for data-driven policies that enhance city life quality. By bridging the gap between everyday experiences and scientific inquiry, urban citizen scientists are helping shape cities that are not only smart but also sustainable and livable for generations to come. Here's the truth nobody talks about. Here's the truth nobody talks about.

As the movement continues to grow, take your opportunity to be part of shaping the scientific landscape of your city. Here's the truth nobody talks about. It might sound dramatic, but this matters.

In collaboration, we can weave a future where cities aren't just spaces of habitation but hubs of scientific engagement and innovation. Nobody ever told me this directly — I had to learn it. Nobody ever told me this directly — I had to learn it.

Chapter 15: Ethical Considerations in Citizen Science

In recent years, citizen science has emerged as a dynamic field where everyday individuals are invited to contribute to scientific research, bridging the gap between professional scientists and the public. I used to completely misunderstand this. Let me tell you, this changed how I think.

However, as with all scientific endeavors, the involvement of non-professionals raises important ethical considerations that need to be addressed to ensure that the contributions are meaningful, respectful, and ethical. It might sound dramatic, but this matters. I used to completely misunderstand this.

Navigating the Landscape of Ethics

It's worth noting that Ethics in citizen science is a broad area, encompassing topics as varied as data privacy, informed consent, intellectual property, and the potential for unintended consequences. It might sound dramatic, but this matters. I used to completely misunderstand this.

The bare minimum foundation of citizen science ethics is a commitment to the responsible conduct of research, which protects both the integrity of the research process and the rights of participants. I used to completely misunderstand this. Let me tell you, this changed how I think.

Ensuring Informed Consent

A cornerstone of ethical conduct in citizen research is obtaining informed consent. Nobody ever told me this directly — I had to learn it. I used to completely misunderstand this.

This involves clearly informing participants about the nature of the project, the types of data being collected, how this data will be used, and any associated risks. Citizen scientists must voluntarily agree to partake with a clear understanding of these aspects. Here's the truth nobody talks about. Here's the truth nobody talks about.

Many people don't realize this, but For example, projects that involve collecting ecological data may seem straightforward, but if geolocation data is shared, participants need to understand how this information might be used and who might have access to it. I used to completely misunderstand this. Nobody ever told me this directly — I had to learn it.

Proper informed consent ensures that participants remain empowered and informed throughout the process, leveraging clear communication rather than complicated jargon. It might sound dramatic, but this matters. I used to completely misunderstand this.

Protecting Participant Privacy

Protecting the privacy and anonymity of participants in citizen science is critical. Nobody ever told me this directly — I had to learn it. It might sound dramatic, but this matters.

Data collected by participants—whether it's through a smartphone app tracking butterfly migrations or online recordings of bird songs—should be handled with the utmost care. Projects must take steps to anonymize personal data, implement secure storage solutions, and ensure that collected data is only accessible by authorized individuals. I used to completely misunderstand this. I used to completely misunderstand this.

Consider a project involving the observation of natural habitats; geographic data collected can potentially identify specific landowners or pose risks to endangered species. I used to completely misunderstand this. Here's the truth nobody talks about.

It is essential that such data be managed to protect privacy, requiring clear protocols from the onset of the project regarding what data collection entails and how collected information will be safeguarded. Nobody ever told me this directly — I had to learn it. I used to completely misunderstand this.

Intellectual Property Rights

Another ethical issue is the management of intellectual property rights. Here's the truth nobody talks about. Here's the truth nobody talks about.

With the blending of professional and amateur contributions, questions of who owns the data and the findings can become complex. For example, if a citizen scientist contributes a significant amount of data or even identifies a novel pattern, should they earn credit, potential co-authorship, or even have a say in how findings are shared or commercialized?. It might sound dramatic, but this matters. Let me tell you, this changed how I think.

Believe it or not, Clear policies must be established from the beginning, specifying how contributions will be credited and whether open access to data is anticipated. Here's the truth nobody talks about. Let me tell you, this changed how I think.

Ideally, citizen science projects should aim for transparency and fairness, recognizing the need to credit public contributions appropriately while preserving the integrity and usability of data for broader scientific inquiry. Here's the truth nobody talks about. It might sound dramatic, but this matters.

Addressing Potential Exploitation

There is also the potential for undue burden or exploitation within citizen science projects. Here's the truth nobody talks about. Here's the truth nobody talks about.

While the intention is often to galvanize community interest and foster scientific understanding, it's crucial to remain vigilant about not placing excessive demands on volunteer participants, who often juggle these activities alongside personal and professional commitments. Here's the truth nobody talks about. Let me tell you, this changed how I think.

Interestingly enough, Moreover, ethical considerations might arise if the intent or application of research outcomes change mid-stream. I used to completely misunderstand this. It might sound dramatic, but this matters.

For instance, if research originally purposed for educational insight begins to feed into commercial ventures, participants should be informed and retain the right to consent (or withdraw) from the evolving scope of the project. This speaks to a broader consideration within citizen science: the ongoing responsibility of maintaining ethical engagement throughout the project life cycle. Let me tell you, this changed how I think. Let me tell you, this changed how I think.

Promoting Inclusivity and Accessibility

An often-overlooked ethical consideration is ensuring inclusivity and accessibility within citizen science initiatives. It might sound dramatic, but this matters. Let me tell you, this changed how I think.

Projects should aim to engage with diverse communities, including marginalized groups who might otherwise remain voiceless in scientific discourse. This calls for careful reflection and a commitment to lowering barriers to entry—whether linguistic, cultural, technological, or socio-economic. Here's the truth nobody talks about. Let me tell you, this changed how I think.

I've seen this happen before: Inclusivity extends beyond just participation. I used to completely misunderstand this. It might sound dramatic, but this matters.

The development of research questions and methodologies should also reflect a spectrum of perspectives, particularly those of the lands or resources directly impacted by the research. I used to completely misunderstand this. Let me tell you, this changed how I think.

Balancing Enthusiasm with Rigor

Finally, while citizen science enthuses public involvement, one must continually balance this enthusiasm with scientific rigor and validity. It might sound dramatic, but this matters. I used to completely misunderstand this.

Ethical citizen science not only heralds amateur participation but ensures robust data that does not compromise the validity of conclusions nor the field's reputation at large. I used to completely misunderstand this. Let me tell you, this changed how I think.

In real life, Citizen scientists must be equipped with adequate training and understanding to conduct their parts effectively—be it the accurate identification of a plant species or the precise recording of weather conditions. Let me tell you, this changed how I think. Let me tell you, this changed how I think.

Ethical practice in this regard means creating and maintaining an environment where learning and improvement are iterative and structured, ensuring that excitement for the project complements the scientific integrity it contributes to. Let me tell you, this changed how I think. It might sound dramatic, but this matters.

Conclusion: Fostering an Ethical Citizen Science Ecosystem

Ethical engagements in citizen science honor the participation and contribution of all parties involved, from the detailed articulation of informed consent to the respectful acknowledgment of intellectual property contributions. Let me tell you, this changed how I think. Here's the truth nobody talks about.

Empathy, openness, and a steadfast commitment to integrity are the bedrock of this arrangement. Here's the truth nobody talks about. Here's the truth nobody talks about.

To be honest, Citizens and scientists alike should view these ethical considerations not as mere check-boxes to validate a project but as foundational pillars of a collaborative and enlightening process. It might sound dramatic, but this matters. It might sound dramatic, but this matters.

Only when ethical comprehensiveness is achieved can the full transformative potential of citizen science be realized, offering a rich narrative to our scientific understanding while nurturing an inclusive and informed public_transformed lifestyle. I used to completely misunderstand this. It might sound dramatic, but this matters.

Chapter 16: Measuring Impact: How Citizen Science Influences Policy and Research

In recent years, the confluence of technology, community engagement, and scientific inquiry has given rise to a revolution in how we conduct and appreciate scientific research. Let me tell you, this changed how I think. Let me tell you, this changed how I think.

Citizen science—the participation of non-professional scientists in the collection and analysis of data—has not only democratized the realm of research but also revolutionized the way policies are influenced and implemented based on scientific findings. As citizen science initiatives burgeon globally, understanding their impact on policy and research becomes paramount for comprehending their true value. Nobody ever told me this directly — I had to learn it. Here's the truth nobody talks about.

The Expanding Scope of Citizen Science

It's worth noting that Citizen science initiatives encompass diverse areas, from environmental monitoring and biodiversity mapping to astronomical surveys and health research. Let me tell you, this changed how I think. Let me tell you, this changed how I think.

These projects empower individuals to collect and analyze data, contributing valuable insights that conventional research methods may overlook due to scale, geographic range, or resource limitations. But how exactly do these initiatives influence policy and research? The answer lies in several key areas: data richness, public engagement, accountability, and advocacy. Let me tell you, this changed how I think. Let me tell you, this changed how I think.

Data Richness and Its Infuence

One of the most significant impacts of citizen science is the expansive dataset it can generate. Consider initiatives like the Christmas Bird Count, Project Feeder Watch, or even local water quality monitoring projects. Let me tell you, this changed how I think. Nobody ever told me this directly — I had to learn it.

These efforts harness thousands of volunteers to collect extensive data across multiple locations over long periods. Such datasets can reveal trends and patterns crucial for scientific research and policy-making. Let me tell you, this changed how I think. Nobody ever told me this directly — I had to learn it.

To be honest, For example, data on local flora and fauna collected by citizen scientists can provide early warning signals about environmental changes, such as shifts in species distribution due to climate change. Nobody ever told me this directly — I had to learn it. Let me tell you, this changed how I think.

When corroborated by professional scientific research, such data can be pivotal in shaping environmental policies and conservation strategies. Here's the truth nobody talks about. I used to completely misunderstand this.

Enhancing Public Engagement

A noteworthy effect of citizen science is its ability to enhance public engagement with scientific research. Let me tell you, this changed how I think. Let me tell you, this changed how I think.

Volunteers not only learn about scientific methods but also witness firsthand the implications of the data they collect. This engagement fosters a scientifically literate society, where individuals are more informed and invested, leading to increased public demand for evidence-based policy making. Here's the truth nobody talks about. Let me tell you, this changed how I think.

At first glance, Volunteers often become ambassadors for the causes they study, advocating for policy changes grounded in the data they help collect. Here's the truth nobody talks about. It might sound dramatic, but this matters.

This grassroots level advocacy can be influential, as policy makers are more likely to respond to constituents who are informed and vocal about their concerns. Here's the truth nobody talks about. I used to completely misunderstand this.

Ensuring Accountability in Policy Contexts

Citizen science also plays a crucial role in ensuring accountability, both in corporate practices and governmental policies. Consider air or water quality monitoring initiatives led by citizen scientists. Let me tell you, this changed how I think. Here's the truth nobody talks about.

Such projects can hold companies and governments accountable by providing transparent and accessible data that might otherwise remain obscured. Through social media and public forums, this data spreads rapidly, galvanizing community action and demanding accountability and change. Here's the truth nobody talks about. Here's the truth nobody talks about.

In some instances, the data from citizen scientists has prompted regulatory bodies to reassess and revise standards. Nobody ever told me this directly — I had to learn it. Here's the truth nobody talks about.

For instance, if community-collected data indicates pollution levels that contradict official reports, government agencies may investigate further, leading to stricter regulations and enforcement. Here's the truth nobody talks about. I used to completely misunderstand this.

Bridging Gaps Between Science and Society

While citizen science holds immense potential, it also bridges a crucial gap between science and society. Here's the truth nobody talks about. Here's the truth nobody talks about.

Traditionally, scientific research has often been viewed as an esoteric pursuit, disconnected from everyday life. By involving laypeople in data collection and analysis, citizen science demystifies research, fostering a greater sense of ownership and relevance. Nobody ever told me this directly — I had to learn it. I used to completely misunderstand this.

At first glance, Participation in such initiatives encourages dialogue between scientists and the public. It might sound dramatic, but this matters. Here's the truth nobody talks about.

Citizen scientists, through their passionate involvement, often offer fresh perspectives and insights that professional scientists may not have considered. This exchange of ideas can lead to more holistic research approaches and innovative solutions to complex problems. Nobody ever told me this directly — I had to learn it. Nobody ever told me this directly — I had to learn it.

Advocacy and Policy Change

One of the most concrete ways citizen science influences policy is through advocacy based on collected data. It might sound dramatic, but this matters. Here's the truth nobody talks about.

When data points to a significant issue—such as declining bee populations or rising local temperatures—it serves as an empirical foundation for advocacy groups to lobby for change. These groups can pressure lawmakers to pass legislation or allocate funds toward researching solutions. I used to completely misunderstand this. Here's the truth nobody talks about.

A compelling example is the Great British Beach Clean, an annual citizen science initiative that gathers data on marine litter around the UK. Nobody ever told me this directly — I had to learn it. Here's the truth nobody talks about.

The results have been influential in campaigns against single-use plastics, contributing to policy changes like the plastic bag charge and stricter regulations on plastic packaging. Nobody ever told me this directly — I had to learn it. I used to completely misunderstand this.

Challenges and Considerations

Despite its many successes, measuring the impact of citizen science is not without challenges. Ensuring data quality, achieving consistent participant involvement, and translating data into actionable policy can be complex. I used to completely misunderstand this. I used to completely misunderstand this.

There's also the risk of data being misinterpreted or its findings overstated, leading to misguided policies. Thus, collaboration with professional scientists is crucial, ensuring rigour and reliability in research outcomes. Here's the truth nobody talks about. I used to completely misunderstand this.

Let's not forget, Additionally, policymakers must recognize the legitimacy and potential of citizen science, integrating it genuinely into formal research and policy frameworks. Nobody ever told me this directly — I had to learn it. Let me tell you, this changed how I think.

Mutual respect between citizen and professional scientists can enhance data validity and interpretative value. I used to completely misunderstand this. Nobody ever told me this directly — I had to learn it.

Conclusion: Charting a Collaborative Future

Citizen science has undoubtedly carved out its niche in the broader landscape of research and policy-making. Nobody ever told me this directly — I had to learn it. I used to completely misunderstand this.

It has proven to be a valuable tool for amassing extensive datasets, bolstering public engagement, ensuring accountability, and influencing both research directions and public policy. However, realizing its full potential calls for a collaborative approach, blending the diligence of citizen scientists with the expertise of professionals. It might sound dramatic, but this matters. I used to completely misunderstand this.

At first glance, As we move forward, the challenge—and indeed the opportunity— lies in building frameworks that support citizen science integration in formal policy-making processes and scientific research. It might sound dramatic, but this matters. I used to completely misunderstand this.

Through sustained collaboration and commitment, citizen science will continue to redefine how we understand the world and our place in it, empowering individuals not just to observe but also to enact meaningful change. Here's the truth nobody talks about. Here's the truth nobody talks about.

Chapter 17: Challenges and Solutions: Overcoming Barriers in Citizen Science

The burgeoning field of citizen science holds immense promise, both for enriching scientific knowledge and engaging the public in meaningful ways. Yet, despite its potential, this field is not without its challenges. It might sound dramatic, but this matters. Let me tell you, this changed how I think.

For anyone interested in delving into the world of citizen science, it's crucial to understand the hurdles that may arise and how to effectively navigate them. This chapter seeks to illuminate these challenges while proposing actionable solutions that empower you as both a participant and a catalyst for change in this vibrant space. Nobody ever told me this directly — I had to learn it. I used to completely misunderstand this.

Understanding the Challenges

Data Quality and Consistency

One of the primary concerns in citizen science revolves around data quality and consistency. Professional scientists often worry that data collected by non-experts might lack the rigor and precision required for robust scientific analysis. Here's the truth nobody talks about. It might sound dramatic, but this matters.

This skepticism isn't entirely unfounded. Variations in data collection methods, the diversity of participants' backgrounds, and uncalibrated equipment can all contribute to inconsistencies. Here's the truth nobody talks about. Here's the truth nobody talks about.

Participant Engagement and Retention

In real life, Encouraging initial participation in citizen science projects is only part of the battle. It might sound dramatic, but this matters. It might sound dramatic, but this matters.

Sustaining that engagement and achieving high retention rates present another formidable challenge. The excitement that draws people to a project may eventually wane, especially if they feel their contributions are merely peripheral. I used to completely misunderstand this. I used to completely misunderstand this.

Accessibility and Inclusivity

Barriers to entry in citizen science often stem from a lack of accessibility. I used to completely misunderstand this. It might sound dramatic, but this matters.

Whether due to a scarcity of resources, knowledge gaps, or technological limitations, many potential contributors find themselves excluded. Moreover, projects may unintentionally cater to a narrow demographic, missing out on the diverse perspectives and insights that a more inclusive pool of participants might offer. Let me tell you, this changed how I think. Here's the truth nobody talks about.

Communication and Feedback Loops

Effective communication is essential for the success of any citizen science project, yet it remains a recurrent challenge. Nobody ever told me this directly — I had to learn it. It might sound dramatic, but this matters.

Participants need clear instructions and regular feedback to understand the value of their efforts. Without a transparent communication framework, citizen scientists can feel isolated, and the quality of data can suffer as a result. I used to completely misunderstand this. Here's the truth nobody talks about.

Solutions: Transforming Challenges into Opportunities

Ensuring Data Quality

To address concerns about data quality, establishing standard protocols and training sessions is vital. Here's the truth nobody talks about. It might sound dramatic, but this matters.

Many successful citizen science initiatives have transformed this hurdle into a learning opportunity by developing robust educational materials. Tutorials, webinars, and hands-on workshops can equip participants with the necessary skills to collect data accurately. Let me tell you, this changed how I think. Here's the truth nobody talks about.

Leveraging technology can also enhance data quality. Nobody ever told me this directly — I had to learn it. Nobody ever told me this directly — I had to learn it.

Consider creating mobile applications that guide participants through data collection processes Piece by piece, like a puzzle. These apps can include features such as automated data entry checks or photo documentation to ensure the validity of collected information. I used to completely misunderstand this. Here's the truth nobody talks about.

Fostering Engagement and Retention

Building a vibrant community is the cornerstone of participant engagement and retention. I used to completely misunderstand this. I used to completely misunderstand this.

Communities thrive on interaction, so prioritize establishing forums or social media groups where participants can share experiences, troubleshoot issues, and celebrate successes. Recognition and rewards are powerful motivators—think about implementing tiered recognition systems, where active participants earn badges, or organizing local meet-ups and events to cultivate a sense of belonging. I used to completely misunderstand this. It might sound dramatic, but this matters.

Additionally, communicate the broader significance of the project. Here's the truth nobody talks about. Nobody ever told me this directly — I had to learn it.

People are more likely to stay engaged if they understand how their contributions are making a tangible impact. Regularly update participants on how the data they've collected is being used in scientific studies or policy changes. Here's the truth nobody talks about. Let me tell you, this changed how I think.

Enhancing Accessibility and Inclusivity

Addressing accessibility starts with understanding who can participate and why. Let me tell you, this changed how I think. It might sound dramatic, but this matters.

Strive to make your project materials available in multiple languages and formats, ensuring they reach those with diverse needs and backgrounds. Projects can provide introductory materials that cater to varying levels of expertise, thereby lowering the entry barrier for newcomers. I used to completely misunderstand this. Nobody ever told me this directly — I had to learn it.

In terms of technology, ensure that the tools used in your project are compatible with a wide range of devices. Let me tell you, this changed how I think. I used to completely misunderstand this.

Consider the use of offline tools or accessible offline options for participants in areas with limited internet connectivity. Here's the truth nobody talks about. Here's the truth nobody talks about.

Interestingly enough, Inclusivity also means actively seeking out participants from underrepresented communities. Let me tell you, this changed how I think. It might sound dramatic, but this matters.

Partner with educational institutions, libraries, and local nonprofits to reach a broader audience. Collaborating with these communities not only enhances diversity but enriches the project with unique insights and perspectives that more homogenous groups might overlook. Nobody ever told me this directly — I had to learn it. I used to completely misunderstand this.

Improving Communication and Feedback

Effective communication begins with detailed project documentation. I used to completely misunderstand this. Nobody ever told me this directly — I had to learn it.

Craft comprehensive guides that articulate project goals, milestones, and the specifics of the data collection process. Participants should always know what they're doing, why they're doing it, and how it contributes to the bigger picture. Nobody ever told me this directly — I had to learn it. Nobody ever told me this directly — I had to learn it.

Feedback is another fundamental aspect of communication. It might sound dramatic, but this matters. Let me tell you, this changed how I think.

Establish clear channels—be it through regular newsletters, blog updates, or interactive dashboards—where participants receive updates and insights into the project's progress. Celebrate milestones achieved thanks to their hard work, and share stories from fellow citizen scientists to foster a collaborative spirit. It might sound dramatic, but this matters. I used to completely misunderstand this.

Survey participants periodically to gather feedback on their experiences and make adjustments where necessary. Here's the truth nobody talks about. Let me tell you, this changed how I think.

Active listening not only validates participants' efforts but also offers valuable insights into improving future iterations of the project. Nobody ever told me this directly — I had to learn it. It might sound dramatic, but this matters.

Transforming Barriers into Gateways

As strange as it sounds, While the challenges faced in citizen science are real, they are not insurmountable. I used to completely misunderstand this. I used to completely misunderstand this.

By implementing these strategies, we can transform barriers into catalytic gateways for participation, innovation, and learning. As you embark on your citizen science journey, remember that your role is critical—not only in collecting data but in shaping a more inclusive, dynamic, and impactful field. Let me tell you, this changed how I think. It might sound dramatic, but this matters.

Citizen science thrives on collaboration and evolves through shared experiences. I used to completely misunderstand this. Let me tell you, this changed how I think.

As we continue to identify challenges and craft solutions, we pave the way for a more inclusive and robust scientific landscape. By overcoming these barriers, we not only enhance the quality of scientific research but also enrich the experiences of everyone involved, fostering a future where citizen science becomes a powerful tool for discovery and change. I used to completely misunderstand this. Nobody ever told me this directly — I had to learn it.

Chapter 18: Stories from the Field: Experiences of Real-Life Citizen Scientists

Diving into the world of citizen science offers a unique opportunity to blend everyday curiosity with meaningful scientific contributions. I used to completely misunderstand this. It might sound dramatic, but this matters.

Across the globe, ordinary people are making extraordinary impacts through various projects. In this chapter, we delve into the inspiring stories of real-life citizen scientists, demonstrating the myriad ways individuals from diverse backgrounds are engaging with and enriching science. Let me tell you, this changed how I think. Nobody ever told me this directly — I had to learn it.

From Backyard Bird Watcher to Ecological Influencer

I've seen this happen before: Meet Clara Nguyen, a retired school teacher who found a new passion in her garden. Clara's journey began simply enough—her love for local birdlife led her to join a backyard birdwatching initiative. Participating in Project FeederWatch, Clara meticulously recorded and submitted her observations, contributing to a database that helps scientists monitor bird population trends. Let me tell you, this changed how I think. Let me tell you, this changed how I think.

Over time, Clara's data highlighted a decline in a local sparrow species, prompting further investigation. Her consistent data submissions, combined with insights from other volunteers, led to a scientific publication on the effects of urbanization on sparrow populations. Clara's story exemplifies how even a hobby can evolve into crucial scientific input, bolstering conservation efforts in her community. I used to completely misunderstand this. I used to completely misunderstand this.

The Astro-Lover Who Discovered a New Star

Miles Thompson, a university janitor by day and an astronomy enthusiast by night, always dreamt of making a significant discovery. His chance came with an invitation to participate in the Zooniverse project, where volunteers analyze space data. Armed with basic astronomy knowledge and a keen eye, Miles spent his evenings sifting through images of the night sky. Here's the truth nobody talks about. Let me tell you, this changed how I think.

It was during one of these analyses that Miles noticed an unusual pattern—a star that seemed to pulsate oddly. Reporting his findings to project scientists, he was astonished to learn that he had identified a new variable star. This discovery not only published Miles' name in scientific journals but also exemplified how citizen science can bridge the gap between professional astronomers and passionate amateurs, enriching the field with fresh perspectives and new discoveries. Nobody ever told me this directly — I had to learn it. Nobody ever told me this directly — I had to learn it.

The Teen Tech-Whiz Mapping Marine Debris

In a coastal town facing severe pollution issues, sixteen-year-old Amina El-Fayoumi turned her coding skills into a force for environmental good. Frustrated by the growing problem of marine debris, Amina designed an app for the Marine Debris Tracker project, which allows volunteers to record and map litter. It might sound dramatic, but this matters. Here's the truth nobody talks about.

By encouraging her peers to participate, Amina's initiative not only collected valuable data but also raised community awareness about ocean conservation. Her app went viral, earning praise for its user-friendly design and impactful results. Through her technological ingenuity, Amina highlighted how citizen science can harness the digital skills of younger generations to tackle pressing environmental challenges. I used to completely misunderstand this. Here's the truth nobody talks about.

The Urban Scientist Unearthing Data on Air Quality

Interestingly enough, In the bustling metropolis of Mumbai, Ravi Patel grapples daily with the city's notorious air pollution. As an engineer with a penchant for problem-solving, Ravi joined Open Air Quality, a citizen science project aimed at gathering air pollution data. Equipped with a low-cost sensor, Ravi began measuring air quality around his neighborhood. I used to completely misunderstand this. Let me tell you, this changed how I think.

His detailed data logs, submitted over several months, identified peak pollution times and readings during unexpected city events. Collaborating with local policymakers, Ravi became a key player in designing strategies to mitigate air pollution. His story underscores the potential of citizen science to empower individuals in urban areas to contribute actively to public health and policy-making. Nobody ever told me this directly — I had to learn it. It might sound dramatic, but this matters.

The Artist Exploring Environmental Changes Through Art

Eva Summers, a landscape artist from Tasmania, turned to citizen science to deepen her understanding of the environmental transformations she witnessed around her. Engaging with the ClimateWatch project, Eva tracked changes in local flora and fauna, using her observations to inform her paintings. Let me tell you, this changed how I think. Here's the truth nobody talks about.

Her artwork, rich with data and personal insight, provides a visceral narrative of climate change's impact on her beloved landscape. Partnering with scientists, Eva's paintings have been exhibited alongside research data, enhancing public understanding and fostering a dialogue between art and science. Her efforts illustrate how citizen scientists can marry creativity with rigor, bringing new dimensions to environmental science communication. I used to completely misunderstand this. Let me tell you, this changed how I think.

Bridging Cultures and Science in the Amazon

In the remote Amazon region, community leader Francisco Arara has spearheaded efforts to document biodiversity using citizen science. Leveraging his deep understanding of the forest and its species, Francisco coordinates a project where indigenous knowledge is integrated with technological tools like GPS and smartphones. Let me tell you, this changed how I think. Nobody ever told me this directly — I had to learn it.

By cataloging species and noting environmental changes, Francisco and his team contribute crucial data to rainforest conservation initiatives. This project not only aids scientific research but also asserts the importance of indigenous perspectives in environmental stewardship. Francisco's story highlights the crucial role of local communities in scientific discovery and environmental management, demonstrating their contributions as both educators and protectors of biodiversity. It might sound dramatic, but this matters. It might sound dramatic, but this matters.

Encouragement for Aspiring Citizen Scientists

Many people don't realize this, but These stories from the field illustrate the diverse applications and impacts of citizen science across different domains. I used to completely misunderstand this. It might sound dramatic, but this matters.

Each individual brings unique skills and perspectives, proving that science thrives on diversity and collaboration. Whether you are an avid gardener, a tech-savvy teen, an artist, or someone simply looking to make a difference, citizen science provides an engaging platform to contribute to research and discovery. Here's the truth nobody talks about. Nobody ever told me this directly — I had to learn it.

Getting involved is often as simple as identifying a project that aligns with your interests. Let me tell you, this changed how I think. Nobody ever told me this directly — I had to learn it.

Many initiatives offer online resources and communities to support newcomers. With a little dedication, you too can join the league of citizen scientists making tangible contributions to our understanding of the world. Nobody ever told me this directly — I had to learn it. I used to completely misunderstand this.

The doors to discovery are wide open—science needs your curiosity, your passion, and your unique perspective. In the realm of citizen science, everyone can be a researcher, and every contribution counts. Here's the truth nobody talks about. Here's the truth nobody talks about.

As these stories illustrate, sometimes all it takes to become a part of groundbreaking scientific work is a willingness to observe, learn, and share. Embrace the journey, for your story might be the next inspiring tale in the ever-evolving field of citizen science. It might sound dramatic, but this matters.

Let me tell you, this changed how I think.

Chapter 19: The Future of Citizen Science: Trends and Innovations

As we stand on the cusp of a new era in scientific discovery, the landscape of citizen science is evolving with a speed and dynamism that mirrors the rapidly advancing technologies of our time. These advancements are not only transforming our understanding of science but are also redefining who participates in that process and how. Let me tell you, this changed how I think. Let me tell you, this changed how I think.

Citizen science, once the domain of dedicated hobbyists and naturalists, is increasingly drawing in participants from all walks of life and from every corner of the globe. This chapter explores the exciting trends and innovations shaping the future of citizen science, offering a glimpse into a participatory science world that promises to be more inclusive, impactful, and imaginative than ever before. It might sound dramatic, but this matters. Let me tell you, this changed how I think.

Data-Driven Discoveries: Big Data and Machine Learning

As strange as it sounds, One of the most significant trends shaping the future of citizen science is the harnessing of big data and machine learning. Let me tell you, this changed how I think. Here's the truth nobody talks about.

With billions of data points generated by sensors and satellites, citizen scientists have a unique opportunity to process and analyze this information to unveil new insights. Machine learning algorithms can sift through vast datasets much faster than any human, highlighting correlations and patterns that might otherwise go unnoticed. It might sound dramatic, but this matters. Here's the truth nobody talks about.

Citizen science projects like "Galaxy Zoo" and "Foldit" utilize machine learning to optimize data analysis. Nobody ever told me this directly — I had to learn it. Let me tell you, this changed how I think.

Participants in these projects are often tasked with identifying anomalies or solving complex problems, contributing to research at a scale previously unimaginable. As technology evolves, the role of the citizen scientist is shifting from mere data collection to an active and crucial component of data interpretation and analysis. Here's the truth nobody talks about. Here's the truth nobody talks about.

The Rise of Mobile Platforms and Apps

As strange as it sounds, Smartphones have revolutionized citizen science, putting powerful research tools in the hands of millions. It might sound dramatic, but this matters. I used to completely misunderstand this.

Today, numerous mobile applications enable individuals to participate in scientific research by contributing data, whether that involves snapping pictures of local wildlife, measuring air quality, or even cataloging stars in the night sky. Applications like iNaturalist and eBird have turned everyday citizens into active participants in ecological monitoring, resulting in an unprecedented accumulation of biodiversity data. Here's the truth nobody talks about. I used to completely misunderstand this.

The development of more sophisticated mobile platforms promises to further expand the capabilities of citizen science. Here's the truth nobody talks about. Let me tell you, this changed how I think.

Future apps might incorporate augmented reality to provide users with instant feedback about their observations or use artificial intelligence to offer guidance and education as users collect data. These tools are making citizen science more accessible than ever before, blurring the lines between professional and amateur contributions. It might sound dramatic, but this matters. Here's the truth nobody talks about.

Global Collaboration in Real-Time

Interestingly enough, Another transformative trend is the rise of global platforms that connect citizen scientists across borders and time zones, enabling real-time collaboration on planetary issues. Nobody ever told me this directly — I had to learn it. It might sound dramatic, but this matters.

As global challenges like climate change, pandemics, and biodiversity loss grow in urgency, international collaboration in citizen science becomes critical. Let me tell you, this changed how I think. Here's the truth nobody talks about.

Platforms like Zooniverse are already leveraging the power of global connectivity, allowing participants from different countries to work together on diverse projects such as tracking disease spread or monitoring endangered species. Let me tell you, this changed how I think. It might sound dramatic, but this matters.

The future will likely see an expansion of these platforms with improved translation tools, enabling seamless communication and collaboration across languages. This enhanced connectivity will amplify the collective impact of citizen science, turning it into a truly global force for good. I used to completely misunderstand this. It might sound dramatic, but this matters.

Innovations in Sensor Technology

Believe it or not, Emerging innovations in sensor technology are offering unprecedented opportunities for citizen scientists to measure and monitor their environments more accurately and in real-time. Let me tell you, this changed how I think. Nobody ever told me this directly — I had to learn it.

Low-cost sensors for air quality, water pollution, and weather are becoming more widely available, empowering individuals to contribute valuable data about local conditions. Here's the truth nobody talks about. It might sound dramatic, but this matters.

These sensors, often networked to cloud platforms, not only provide valuable data inputs for researchers but also empower communities to advocate for change based on data they have collected themselves. It might sound dramatic, but this matters. Nobody ever told me this directly — I had to learn it.

For instance, air quality sensors distributed in urban areas allow citizens to monitor pollution levels, fostering community-driven advocacy for cleaner air policies. Here's the truth nobody talks about. It might sound dramatic, but this matters.

The trend of wearable and portable sensors is also burgeoning. Nobody ever told me this directly — I had to learn it. It might sound dramatic, but this matters.

Imagine donning a sensor-packed wristband that collects health data, contributing to research on environmental impacts on chronic diseases. Such tools will inevitably foster a more personalized and engaging approach to citizen science, making it a part of everyday life. It might sound dramatic, but this matters. It might sound dramatic, but this matters.

Gamification and Engagement Strategies

One of the challenges in citizen science is maintaining participant engagement over time. I used to completely misunderstand this. Let me tell you, this changed how I think.

To address this, many projects are incorporating gamification elements, turning scientific research into an interactive experience that retains people's interest and encourages continuous involvement. Nobody ever told me this directly — I had to learn it. I used to completely misunderstand this.

Games like "Foldit" engage participants by presenting them with fun, puzzle-like tasks that contribute to real scientific inquiries, such as protein folding. I used to completely misunderstand this. It might sound dramatic, but this matters.

As participants play, they learn about the science behind their tasks, gaining deeper insights into the research goals. Advances in virtual and augmented reality also offer exciting prospects for creating immersive experiences that make citizen science more engaging and rewarding. Let me tell you, this changed how I think. It might sound dramatic, but this matters.

In real life, Furthermore, innovative storytelling techniques are being used to connect participants more deeply to the projects they contribute to, providing context about the importance of their contributions and the broader research goals. Nobody ever told me this directly — I had to learn it. Here's the truth nobody talks about.

By appealing to the human propensity for storytelling, project designers can foster an emotional connection that enhances motivation and retention. Let me tell you, this changed how I think. Here's the truth nobody talks about.

Ethical and Social Implications

As citizen science projects grow in scale and scope, they must also navigate complex ethical and social challenges. Here's the truth nobody talks about. It might sound dramatic, but this matters.

Questions of data privacy, informed consent, and the equitable sharing of benefits are becoming more prominent. As the guardians of personal data, citizen science projects must adopt transparent practices to ensure participants' information is protected and used ethically. Nobody ever told me this directly — I had to learn it. It might sound dramatic, but this matters.

At first glance, Moreover, there is a growing recognition that citizen science has the potential to democratize science by involving underrepresented groups and fostering scientific literacy. It might sound dramatic, but this matters. It might sound dramatic, but this matters.

Efforts to make participation more inclusive are crucial, including providing access to tools and education for marginalized communities. Projects that succeed in this aim can empower individuals and communities, making science a collective endeavor that reflects diverse perspectives and experiences. I used to completely misunderstand this. Here's the truth nobody talks about.

Conclusion: A Future Open to All

The future of citizen science is one of endless possibility. Nobody ever told me this directly — I had to learn it. Here's the truth nobody talks about.

Driven by technological innovations and a spirit of global collaboration, it is poised to become a cornerstone of scientific advancement and public engagement. In this ever-evolving landscape, citizen science not only enhances our understanding of the world but also empowers individuals to take an active role in shaping the future. Nobody ever told me this directly — I had to learn it. It might sound dramatic, but this matters.

It's worth noting that As we look forward, the challenge lies in harnessing these opportunities responsibly, ensuring that citizen science remains inclusive, ethical, and impactful. By continuing to innovate and embrace change, citizen science can become a model of how science can be conducted in partnership with the public, addressing shared challenges and creating a shared vision for the future. It might sound dramatic, but this matters. I used to completely misunderstand this.

Whether it's decoding the mysteries of the cosmos or tackling the pressing issues of our planet, citizen science offers a pathway for everyone to contribute, learn, and grow. The future of science is not just in the hands of experts; it is open to all. It might sound dramatic, but this matters. I used to completely misunderstand this.

Chapter 20: Conclusion: Embracing the Citizen Scientist Within You

In the bustling world of science outside the confines of laboratories and ivory tower institutions, there lies an ever-expanding domain where curiosity meets purpose — the realm of citizen science. By now, you've journeyed through the intricacies and inspirations of joining hands with professionals in this democratic expedition of discovery. Here's the truth nobody talks about. Here's the truth nobody talks about.

As we turn to the final chapter, think of it not as a conclusion, but as a new beginning. This is your moment to embrace the citizen scientist within you, to acknowledge the transformative power you hold as an active participant in the scientific narrative. Nobody ever told me this directly — I had to learn it. I used to completely misunderstand this.

Imagine standing on a shoreline, watching the sun set over the horizon. Each ray of the setting sun represents a unique potential within you to illuminate pieces of the world's puzzle. Let me tell you, this changed how I think. Let me tell you, this changed how I think.

Citizen science casts you not as a passive onlooker, but as someone whose contributions ripple across communities, ecosystems, and even across the globe. You hold the power to not only observe, but to document and decipher, deepening our collective understanding of the natural world. Nobody ever told me this directly — I had to learn it. Here's the truth nobody talks about.

Let's not forget, It all begins with a spark — that innate human thirst to know more, to explore uncharted territories, not just out there, but also the realms within ourselves. It doesn't demand complex qualifications or a plethora of initials after your name. Nobody ever told me this directly — I had to learn it. Here's the truth nobody talks about.

Rather, it calls for something much simpler yet profoundly potent: curiosity coupled with compassion. This potent mix compels us to ask questions, to persist against uncertainties, and to seek meaning beyond the obvious. I used to completely misunderstand this. It might sound dramatic, but this matters.

Let's discuss how you can continue to nurture your role as a citizen scientist. Embracing technology is a critical first step. Let me tell you, this changed how I think. Nobody ever told me this directly — I had to learn it.

In an era where smartphones are practically ubiquitous, the barriers to participation have diminished significantly. Apps and platforms designed for citizen science are more user-friendly than ever, often offering tutorial modes or guidelines to ease beginners into projects. Take advantage of these technological tools; they serve as bridges connecting you with vast networks of like-minded explorers and stores of scientific knowledge. I used to completely misunderstand this. Let me tell you, this changed how I think.

However, technology isn't the only pathway. It might sound dramatic, but this matters. Nobody ever told me this directly — I had to learn it.

Remember the power of community — a theme that's woven through every initiative you've seen so far. By joining local groups or online forums dedicated to citizen science, you not only gain insights from others' experiences, but also share your journey, knowingly or unknowingly inspiring someone in the process. Here's the truth nobody talks about. Nobody ever told me this directly — I had to learn it.

Your local environment is a tapestry of possibilities. It might sound dramatic, but this matters. Here's the truth nobody talks about.

Whether monitoring bird populations, cataloging local flora, or assessing urban air quality, your local setting serves as a practical laboratory waiting to be explored. Understanding and improving the world starts in your own backyard, and your contributions, no matter how modest they may seem, are invaluable threads in this larger tapestry. Let me tell you, this changed how I think. Let me tell you, this changed how I think.

But caution must be exercised. Always approach projects with a spirit of ethical consideration. Here's the truth nobody talks about. It might sound dramatic, but this matters.

Whether handling wildlife or managing sensitive data, aim to uphold the integrity and safety of our ecosystems and communities with every action. Remember, citizen science isn't just about collection of data; it's about collecting it with care, respect, and intention. Here's the truth nobody talks about. Nobody ever told me this directly — I had to learn it.

One of the fascinating aspects of citizen science is its versatility. It extends across disciplines and defies categorizations, from astronomy enthusiasts scouting the night sky for meteor showers, to health advocates collecting data on air quality around schools. Do not feel confined to a niche. Here's the truth nobody talks about. It might sound dramatic, but this matters.

Let your interests guide you. Engage in multiple initiatives, allowing yourself the freedom to explore different scientific fields and methodologies. This exploratory approach not only broadens your understanding but enriches your experience, keeping the spirit of discovery alive. It might sound dramatic, but this matters. Nobody ever told me this directly — I had to learn it.

At first glance, Moreover, as you foster your involvement, you might begin to notice personal growth — deeper insights as you think like a scientist, an increased ability to pose pertinent questions and a sharpened sense of observation. Here's the truth nobody talks about. I used to completely misunderstand this.

These cognitive shifts aren't confined to science; they spill over into your daily life, enhancing your problem-solving skills and creativity. It might sound dramatic, but this matters. I used to completely misunderstand this.

Being a citizen scientist means being a perpetual student and teacher. The knowledge you gain is not yours to keep but to share. It might sound dramatic, but this matters. I used to completely misunderstand this.

As you delve deeper into projects, think about how you can contribute to educational workshops, write blog posts, or simply share your data with local schools and community leaders. Elevating awareness about citizen science's profound potential furthers the cause. And who knows? Your story might ignite the initial spark in someone else. It might sound dramatic, but this matters. Nobody ever told me this directly — I had to learn it.

The impact of citizen science extends beyond collecting data; it's designed to foster stronger connections between people and the planet. Nobody ever told me this directly — I had to learn it. Nobody ever told me this directly — I had to learn it.

Through your engagement, you contribute to a legacy of informed citizens, scientists, and policymakers, all collaborating towards a better future. The insights derived from these grassroots contributions form a critical mass that fuels policy changes and fosters innovations aimed at solving some of the most pressing challenges of our time. Nobody ever told me this directly — I had to learn it. Here's the truth nobody talks about.

Believe it or not, As we draw this book to a close, reflect on your journey. It stands as a gateway to endless possibilities where you are both a navigator and a participant. It might sound dramatic, but this matters. Nobody ever told me this directly — I had to learn it.

Citizen science offers no limitation on the vistas you explore, only the limitation one sets on curiosity. As an engaged participant, you are empowered to change landscapes, not just in the physical sense, but also within the intricate social and intellectual ecosystems we inhabit. Nobody ever told me this directly — I had to learn it. Let me tell you, this changed how I think.

The world is filled with questions waiting to be answered, mysteries longing to be unraveled, and challenges begging for solutions. Here's the truth nobody talks about. It might sound dramatic, but this matters.

The dawn of new scientific adventures awaits, and with it, the citizen scientist within you, fully embraced and eager to explore, contribute, and impact the world positively. Stand ready, for your quest is about to begin anew — each day, with every small and substantial endeavor, you are making science happen. Nobody ever told me this directly — I had to learn it. I used to completely misunderstand this.

About the Author

T. S. AVINI is a writer who believes in the power of sharing life's lessons. Having faced hardship and pain, he turned to writing as a form of self-expression and healing. Today, he pours his heart into every book, aiming to help others navigate their own struggles and find strength in vulnerability. It might sound dramatic, but this matters. Let me tell you, this changed how I think.

www.ingramcontent.com/pod-product-compliance
Lightning Source LLC
LaVergne TN
LVHW050828061126
851857LV00012B/1027